EMERGENCE AND EVOLUTION OF ENDOGENOUS WATER INSTITUTIONS IN AN AFRICAN RIVER BASIN

Local Water Governance and State Intervention in the Pangani River Basin, Tanzania

CRC Press
Taylor & Francis Group
Boca Raton London New York

CRC Press is an imprint of the
Taylor & Francis Group, an **informa** business

A BALKEMA BOOK

To Dorothy Atek, my mama
Hopefully this may be considered a contribution towards your unfulfilled wish
to go to school!

EMERGENCE AND EVOLUTION OF ENDOGENOUS WATER INSTITUTIONS IN AN AFRICAN RIVER BASIN

Local Water Governance and State Intervention in the Pangani River Basin, Tanzania

DISSERTATION

Submitted in fulfilment of the requirements of
the Board for Doctorates of Delft University of Technology
and of the Academic Board of the UNESCO-IHE Institute for Water Education
for the Degree of DOCTOR
to be defended in public
on Thursday, July 4, 2013, at 12:30 hours
in Delft, The Netherlands

by

Charles Hans KOMAKECH

MSc Water Management, UNESCO-IHE, Delft, The Netherlands
MSc Water and Waste Engineering, Loughborough, United Kingdom
BSc Civil Engineering, Makerere University, Kampala, Uganda
born in Kitgum, Uganda

CRC Press
Taylor & Francis Group
Boca Raton London New York

CRC Press is an imprint of the
Taylor & Francis Group, an **informa** business

A BALKEMA BOOK

This dissertation has been approved by the promotor:
Prof.dr.ir. P. van der Zaag

Composition of the Doctoral Committee:

Chairman:	Rector Magnificus, TU Delft
Vice-chairman:	Rector, UNESCO-IHE
Prof.dr.ir. P. van der Zaag	UNESCO-IHE/ TU Delft, promotor
dr.ir. B. van Koppen	International Water Management Instute, Sri Lanka*
Prof.dr. B. Lankford	University of East Anglia, United Kingdom
Prof.dr. F. Cleaver	King's College London, United Kingdom
Prof.dr.ir. H.H.G. Savenije	TU Delft
dr.ir. J.A. Bolding	Wageningen University
Prof.dr.ir. N.C. van de Giesen	TU Delft, reserve member

*dr.ir. B. van Koppen has provided substantial guidance and support in the preparation of this thesis.

CRC Press/Balkema is an imprint of the Taylor & Francis Group, an *informa* business

© 2013, Charles Hans Komakech

Published by:

CRC Press/Balkema
PO Box 11320, 2301 EH Leiden, The Netherlands
e-mail: Pub.NL@taylorandfrancis.com
www.crcpress.com – www.taylorandfrancis.co.uk

ISBN 978-1-138-00111-4 (pbk)

ABSTRACT

Water management challenges in basins of Sub-Saharan Africa and in other parts of the world are increasing due to rapid urbanisation, poverty and food insecurity, energy demands, and climate change. Nearly half of the world population live in cities, and this is estimated to reach two-thirds of the world's population by the year 2050. The need to improve water services in cities poses new challenges to river basin management. Water transfer from other sectors to cities is an obvious way of reallocating the uses and users of the available water but this may have far reaching upstream-downstream consequences in a catchment. In addition there is an increasing trend in rural poverty, hunger, and food insecurity in Sub-Saharan Africa. To reduce and/or reverse the increasing trend of rural poverty and generate employment requires substantial investment in irrigated agriculture in Sub-Saharan Africa. However, transforming Sub-Saharan Africa's agriculture also implies intervention in water control as lack of access to reliable water supply is one of the major limitation to crop production. Coupled with the above problems are the rising global food and energy prices which have attracted foreign investment in agricultural land in Sub-Saharan Africa. Foreign direct investment in agriculture in Sub-Saharan Africa is likely to increase agricultural water use and this could lead to further enhancement of an already stressed water situation.

In many places the users as well as the State attempt to respond to the challenges, by diverting more water from the river, by building storage reservoirs or by looking for alternative water sources (groundwater use). These responses are likely to cause water scarcity thereby affecting users in other parts of the basin. Increased water scarcity leads to competition and conflict between users, large and small, up and downstream. The increasing competition over water puts additional demands on existing water institutions, and their capacity to reconcile competing claims. In addition to supply augmentation measures, solving water competition and conflict requires crafting new governance arrangements that can ensure equitable and sustainable use of the limited water resources. These include devising rules of how water is shared among competing users and the institutional arrangements to monitor and ensure compliance with the allocation mechanism. This makes understanding the processes of institution change and implementation approaches central to solving water management challenges faced by society in water stress catchments across the globe.

Many governments in Sub-Saharan Africa have adopted new policies and laws, and established new institutions to achieve equitable and sustainable management of water resources. The formalisation of the property right to water and users participation through catchment forums is considered to improve coordination and solve water conflicts. However, government-led water policies and institutional arrangements fail to take local water management practices into account. Some local water management practices are well known historically, especially in (semi)arid regions as these developed into successful institutions for sharing water. Locally evolved governance approaches if well understood could be a substitute for, or used to improve, catchment water institutions being implemented by many governments. The challenge is that adopting this approach implies that local level approaches will be up-scaled while state led institutions are down-scaled. This also requires understanding why local institutional arrangements emerge, and how they function and are being sustained and the scales at which they remain effective.

This thesis contributes to this project by studying one African river basin, namely the Pangani river basin, Tanzania. The basin is a perfect living laboratory to study the emergence and evolution of local and state-led water management institutions. Pangani is a partially closed river basin, partially because some of its tributaries do not flow throughout the year due to over use. It is partially open in that groundwater use is still limited but also because there is very little knowledge on groundwater use, availability, interaction with surface water. It is a basin where state-led intervention dates back from the colonial era and local practices evolved over a period of more than 100 years. The overarching research objective was to explore conditions for reconciling state-led institutional arrangements and local water management practices. This thesis is based on findings from multiple case studies in the Pangani river basin, Tanzania. In-depth interviews, role play games administered through feedback workshops were used to engage in multiple dialogues with the object of research, and all this based on a meticulous cartography of irrigation canals and irrigated plots and zones.

The findings in this thesis indicate that instead of harmony, the states' intervention in the water sector appears to generate dissonance at the interface with locally evolved water institutions. In the Pangani basin state-led formalisation of the property right to water is being used by new actors to gain access and control to water at the expense of existing users. Water rights as implemented in the Pangani river basin are difficult to enforce and control, and so far has not led to efficient water use. There is a problem with enabling meaningful participation by the resource users in decision making related to catchment water management. In one catchment, the Kikuletwa, it proved difficult to define the most appropriate hydrological management unit for decision-making that was able to fit well with the political-administrative territories. The way institutional nesting was done in the Kikuletwa catchment did not work. Modularisation of the larger Kikuletwa catchments into smaller sub-units to form sub-catchment water users associations only created additional water management layers without necessarily integrating locally evolved arrangements such as the river committees. The newly created Kikuletwa sub-catchment water users associations are like islands of associations not well integrated with the existing arrangements. Water users do not see how the sub-catchment water users association is linked to their own governance arrangements. The general conclusion on state intervention in water management is that resolving the problem of institutional fit while integrating customary arrangements with the state-led governance structure requires careful analysis of the existing local structures, and a good understanding of their strengths and limitations.

Although it is widely considered that allocating water rights or use permits would in water stressed catchments improve equity and reduce conflict, the findings in this thesis indicate that the 'paper' based water rights may be used by new actors to gain access to water. The water rights system as administered by the Tanzanian government in the Pangani basin provides the legal means for powerful actors to dispossess existing users. Powerful cities in the Pangani basin selectively used the law to gain leverage over water control. In other cases the legitimacy of the state-based water rights system is questioned by several actors. In the Pangani basin, small scale users appeal to customary principles while large-scale irrigators attempt to gain water access using the state's statutory water law. Although most of the estates have location advantage, their 'official water right' does not go unchallenged by the downstream smallholder farmers. These farmers demand that allocation should be rotational and take into account supply variability and not the absolute values specified in the government water right.

The thesis showed that local level innovation in institutional arrangements for water sharing often emerged around the creation of hydraulic property and/or was negotiated to secure more water flow for downstream users. The hydraulic position of the various actors in a catchment (upstream or downstream) is the main driver for institutional innovation. In the cases studied it was always the downstream users that initiated the process of institutional change in a catchment. Unlike most research on collective action in which water asymmetry, inequality and heterogeneity are seen as risks to collective action, this thesis found that they dynamically interact and give rise to interdependencies between water users which facilitate coordination and collective action. The findings on collective actions are confined to relatively small spatial and social scales, mostly involving irrigators from one village. In such situations there may be inhibitions to unilateral action due to social and peer pressure. Proximity may thus be a necessary condition for collective action in water asymmetrical situations to emerge but at larger spatial scales and over greater distances, for example when considering entire catchment areas or river basins, this is likely to be different. The largest spatial scale where local resource users managed water allocation was a river stretch of 3 administrative wards (spatial distance of about 15.0 km) managed by a local river committee.

This thesis contributes to existing theories and concepts related to catchment water management. The thesis expanded Molle's (2003) typology of basin actors' responses by explicitly introducing a meso layer which depicts the interface where state-led and local-level initiatives and responses are played out. It also showed that not all the eight design principles proposed by Ostrom (1993) are necessary for a water institution to be effective and to endure over time. The thesis also provides conceptual clarity to the dynamics between water asymmetry, inequality in access to land, and heterogeneity sustaining collective action over common pool resources.

In conclusion, local solidarity approaches function best at the scale in which they are currently found, normally involving about 2-3 administrative wards or just a river stretch. No locally created arrangement was found beyond the spatial scale of a river stretch. This is likely because beyond the small spatial scale, they may be difficult to initiate and sustain or they may even collapse. Hierarchical structure that nests local water management arrangements did not work in the catchment studied partly because of the way it was implemented but also due to the complex overlapping jurisdictions between state-led and locally evolved ones. However there is a possibility to integrate state-led river basin management structure with local water management arrangements. In the Pangani basin, we find the river committee as the most promising locally evolved institution that can reconcile state-led and locally created water institutions. As a policy recommendation, a river committee could be issued collective water rights with a mandate to guarantee a minimum amount of water flow downstream of its area of jurisdiction. This way the basin water boards would need not issue water rights that they can not enforce and control, instead they would invest their limited resources to monitor compliance by the river committees.

However, research is needed to understand the role village government can play in addressing competition over water at larger spatial scales. This also thesis did not discuss in-depth the dynamic of gender, inequality and access to water. Leadership of local as well as state-led water management organisations in the Pangani basin are male dominated and in such a situation equity and fairness with respect to gender may be compromised. Better reconciliation of state-led and local water management arrangements with fewer opportunities and better checks for the more powerful to widen inequities may as well benefit women and other marginalized groups; this requires further research. Unlike the Pangani, most basins in Sub-Saharan Africa are still open. In this basins, supply augmentation from alternative sources may still be the first the first step. However, since the creation of hydraulic property also changes the relation between the actors there is a need for further research to compare hydraulic property rights creation (infrastructure development to increase supply) with institutions to share a limited supply. Finally to provide more insight into the functioning of self-governing institutions, further research is needed: 1) to describe phenomena of water asymmetry, inequality and heterogeneity at larger spatial scales, and to analyse under which circumstances they occur; and 2) to verify the relation between inequality of access to land and water in furrow systems and the collective ability to share water and mobilize labour for maintenance at many other furrow systems in order to generalize the findings of this thesis, not only in the Pangani but also in the Rufiji river basin in Tanzania, as well as in other African countries, such as Kenya and Mozambique, and perhaps even in other continents, such as in Nepal.

ACKNOWLEDGEMENT

Quoting Sir Isaac Newton (1676) famous phrase "if I have seen further it is by standing on the shoulders of giants"; I would like to acknowledge that during my research attempt at understanding the emergence and evolution of water institutions I benefited from the works of other scholars in the same field. I would like to acknowledge the constructive support and advice of my supervisor Prof. Pieter van der Zaag and co-supervisor Dr. Barbara van Koppen. I remember after finishing my Master thesis at UNESCO-IHE, Pieter asked if I was interested in a PhD research position in the Pangani. I will remember those scribbles on the sides of my drafts "be precise" or "it is nearly there" and the fun times as well, for instance "the famous safari down the Pangani river". It is still not clear though if it is better to engage 4-wheel to drive upstream of a river than use higher gears without 4-wheel when your car clutch plate fails at the estuary.

I spent three months at Carnegie Mellon University, Pittsburgh learning how to develop agent-based models with Prof. John H. Miller. Thanks John for introducing me to the techniques of game theory and agent-based modelling. The bagel Friday was really good. I am grateful to Prof. Emeritus Francis Clay McMichael for helping me revive my rusty hydraulics and fluid mechanics knowledge. At Carnegie Mellon University I also met the bullpen fellows (PhD students); I would like to thank Sudeep Bhatia, Mark Patterson, Amanda Markey and Nazli Turam for the wonderful time around Pittsburgh.

This research was undertaken as part of the Smallholder System Innovations in Integrated Watershed Management (SSI) programme funded by the Netherlands Foundation for the Advancement of Tropical Research (WOTRO), the Swedish International Development and Cooperation Agency (SIDA), the Netherlands Directorate General for International Cooperation (DGIS), the International Water Management Institute (IWMI) and UNESCO-IHE Institute for Water Education. The Soil-Water Management Research Group at Sokoine University of Agriculture, Tanzania facilitated some of my fieldwork in Makanya catchment. I am thankful to the Pangani Basin Water Office, Moshi, particularly to Hamza Sadiki, now the Director of Water Resources Management, Tanzania for being very supportive of my research. I thank SNV Arusha staff, especially Joel Kalagho and Josephine Lemoyan for making me part of SNV. I appreciate the time of the local water user communities, villages, wards and districts I worked with during the research. I made many friends during my PhD research both in Tanzania and in the Netherlands some of whom also helped with Swahili-English and English-Dutch translations. I can't mention all but thank you Fred Tarimo (RIP), Linus Kiberenge, Raymond Mokiwa, Lily A. Msemo, Angelina Christian, Maliki Abdallah and Ally Hussein. I am also grateful to Vendalin Basso, meneer Ronald Bohté the baridis were wonderful, Lukas Kwezi, Fanuel Karugendo (Mzee wa Pori), mevrouw Danielle Hofboer (mama Nienke), Simone Patzke (mama Juri), Chris de Bont. From Spiritan House I would like to thank Fr. Honest Munish, Sr. Edigna and Sr. Leiticia for hosting me.

I also benefited a lot from the excellent works of my SSI PhD and Post-doc colleagues. Particularly I would like to acknowledge the supports of Keneth Masuki, Siza Tumbo, Elin Enfors thanks for the Suzuki as well, (by the way it is still working if you want it back we can negotiate), Mzee mshauri Hodson Makurira I still remember how to count - hamsini-hamsini but now it is mia, to my friend Marloes Mul thanks for meeting me at the Makanya bus stop - that first night in Makanya village was weird but seeing you walk in the dark was reassuring, Jeltsje Kemerink thanks a lot for the lively discussion, I know you are just about the corner as well, Victor Kongo, Jayashree Pachpute and Line Gordon thanks, Jeremiah Kiptala keep going man. I also would like to thank all the SSI Master students particularly the ones I directly worked with Tulinumpoki Mwakalukwa and Madison Condon.

UNESCO-IHE and the Netherlands at large has always been home away from home. I have made so many friends during my many years of study in the Netherlands. I would like to acknowledge the support of Frank GW Jaspers, who was also my Master thesis mentor, Jolanda Boots, Patricia Davis, Susan Graas thanks for the English-Dutch translations as well (by the way it took five Dutch girls to translate the abstract), my friend Rozemarijn ter Horst and fellow PhD colleagues. Marlou and Pieter apwoyo matek for the many dinner invitations, I enjoyed the groundnut dish! It is still funny to me that you were also in Kitgum at the time I was starting my education journey. After my field research in the Pangani, I became so much interested in giving back to the furrow irrigation farmers. My new job at the Nelson Mandela African Institute of Science and Technology is the right place to do this. I would like to thank Prof. B. Mwamila for offering me the opportunity to put my research knowledge into practical use.

Finally, I would like to thank and acknowledge the support of my family and friends back at home in Uganda. When my sisters first took to me to some distant village school, no one believed them I would ever study. I used to runaway or hide in the bushes; for the seven miles walk was just too far to me. The good news is that since then I have been running to school on my own.

PREFACE

Although some of the water problems can be solved by supply augmentation for instance by building more storage reservoir, rainwater harvesting, looking for alternative sources (groundwater), or investing in green water management, the development of institutions governing use, access and management of the resource is equally important but also the most challenging. There is a general believe that this can be achieved through a social engineering approach where key design principles can be used to craft legitimate arrangements for resource management. However, empirical research shows that institutions emerge and evolve through a process whereby new arrangements are creatively developed using existing way of doing things - bricolage. In addition recent research suggests that incorporating local hydrosolidarity based principles into state-led laws, policies and structures could be a substitute for, or used to improve, catchment water institutions being implemented by many governments. However, it is not clear how to up-scale local arrangements while state-led institutions are down-scaled. The motivation of this PhD research was to identify conditions for reconciling state-led and local water management practices. This also requires understanding why local institutional arrangements emerge, and how they function and are being sustained and the scales at which they remain effective.

After completing my Master thesis research at UNESCO-IHE, Pieter, who was also my supervisor, gave me a flyer and said 'Hans you can submit an abstract to this conference', which I did. In September 2006, I was invited to present at the 3rd International Symposium on Integrated Water Resources Management, Bochum, Germany. I met Pieter at the workshop and shared with him a concept note I was developing for a PhD research. After about a week, Pieter emailed that there was a PhD position in the SSI project in Tanzania but that the budget was small. I didn't really care much about the budget limitation, so I quickly accepted the offer. I became part of the multi-disciplinary SSI research team, most of whom were at the final stages of their field research in the Pangani basin. This was initially a challenge to me, as I would be requested to submit my research findings when I had hardly started field work. It soon became clear that I would be fully on my own in the field as my colleagues finished their field research and graduated.

I found a home in Pangani Basin Water Office and later SNV. Networking with PBWO and SNV made it possible for me to communicate my research findings to farmers and actors interested in the Pangani basin. Through this research, I have discovered to like researching water institutions, agent based modelling and education. I have gained a lot from my research in Pangani basin. Following the water, actors, and thus learning why local hydrosolidarity based institutions emerge, function and evolve over space and time proved worth studying. I must admit though that social science was never a discipline I dreamt about. I remember that at one point in my educational journey, I was admitted to study history, economic and geography at advanced level of secondary school; but I declined and instead studied what I felt was more interesting - physics, chemistry and mathematics. Later I obtained a Bachelor of Science in Civil Engineering. My PhD research, however, allowed me to bridge social sciences and civil engineering. I don't know what name this new profession is, may be a Socio-technical Engineer or may be not. All I know is that I like researching water institutions and I believe that understanding why local institutions emerge, function and evolve over time and space requires skills in both the social and natural sciences. I also learned that it is also possible to gain more insight into the dynamics of water institutions using the techniques of agent-based modelling and participatory gaming with real farmers.

TABLE OF CONTENT

Chapter 8 The role of statutory and local rules in allocating water between large and small-scale irrigators in an African river catchment   167

PART 4: EVOLVING WATER INSTITUTIONS: DISCUSSION AND CONCLUSIONS   189

Chapter 9 A game theoretic analysis of evolution of cooperation in small-scale irrigation canal system   191

Chapter 10 Discussion and conclusions: the emergence and evolution of water institutions   203

LIST OF ACRONYMS

HIBA	Hingilili Irrigation Basin Association
IUCN	The International Union for Conservation of Nature
IWMI	International Water Management Institute
JICA	Japan International Cooperation Agency
NGO	Non-governmental Organisation
PBWB	Pangani Basin Water Board
PBWO	Pangani Basin Water Office
SNV	Netherlands Development Organisation originally established as Stichting Nederlandse Vrijwilligers (Foundation of Netherlands Volunteers)
SSA	Sub-Saharan Africa
SSI	Smallholder Systems Innovations in Integrated Watershed Management
SUA	Sokoine University of Agriculture, Tanzania
TANESCO	Tanzania Electric Supply Company
TIP	Traditional Irrigation Improvement Project
URT	The United Republic of Tanzania
WUA	Water User Associations
WUG	Water User Groups

LIST OF FIGURES

LIST OF TABLES

LIST OF BOXES

PART 1: WATER GOVERNANCE CONTEXT

Water resources in a river basin or catchment have conceptually been distinguished as blue water to include water in rivers, lakes, aquifers, dams and as green water to signify soil moisture (Falkenmark 2007). The process of rainfall partitioning into green and blue water is often influenced by the activities of users located within the river basin or catchment. Here, green water may be used in productive activities such as rainfed agriculture or may provide environmental services, while the blue component could be diverted for irrigation, domestic and industrial purposes and the balance flows downstream where it may be subsequently used or provide environmental services. The blue water flow downstream spatially knit multiple users with different preferences and perceptions of the resource availability in the catchment.

While there is still sufficient water in the river to satisfy all the demands on the water resources, the existence of diverse interests is by itself not a problem. However, with a growing population and other intervening phenomena such as climate change, water resource availability and demands change over time and space. The users as well as the State may respond to the changes, diverting more water from the river, by building storage reservoirs or by looking for alternative water sources. The responses are likely to cause water scarcity thereby affecting users in other parts of the basin. Increased water scarcity leads to competition and conflict between users, large and small, up and downstream. Water conflict may arise when upstream users extract most of the water and leave their downstream neighbours with scarcity. Rising water conflicts and competition described above as well as global processes such as climate change presents management challenges in many river catchments around the world.

The increasing competition over water observed in many river catchments in Sub-Saharan Africa puts additional demands on water institutions, and their capacity to reconcile competing claims. The challenge of river basin governance relates to overcoming coordination problems and defining water institutions able to ensure equitable and sustainable management of the asymmetric common pool resource at various scales and levels.

This thesis consists of four parts, which reflect my journey of discovering water institutions. The first part sets out the great challenges of managing water in an African river catchment where demands outstrip availability (Chapter 1). It highlights concepts and theories that have been recommended to overcome the challenges of water resources management. It briefly discusses concepts and theories such as Integrated Water Resources Management, river basin management, water institutional design principles, institutional bricolage, hydraulic property rights and the problem of collective action institutions for catchment water management. These concepts and theories are confronted with detailed case studies in Parts 2 and 3 of this thesis. Part 1 also introduces the case study area (Chapter 2). Chapter 2 uses the concept of basin development trajectory to illustrate the importance of understanding how local level institutional arrangements interface with national level policies and basin-wide institutions.

Chapter 1

INTRODUCTION

1.1 THE SETTING: WATER MANAGEMENT ISSUES AND CHALLENGES

Water management challenges in basins of Sub-Saharan Africa and in other parts of the world are increasing due to rapid urbanisation, poverty and food insecurity, growing energy demands, and climate change. First, nearly 50% of the world population live in cities, and this is estimated to reach two-thirds of the world's population by the year 2050. The population in Sub-Saharan Africa is projected to be between 1.5 and 2 billion in 2050 and about 50% of which will be living in cities (Faurès and Santini 2008). The growing cities will need a steadily increasing share of the available water resources (de Fraiture and Wichelns 2010). The need to improve water services in cities poses new challenges to river basin management. Transferring water from other sectors to cities is an obvious way of reallocating the uses and users of the available water (Celio et al. 2010) but this may have far reaching upstream-downstream consequences in a catchment.

Second, poverty, hunger, and food insecurity in Sub-Saharan Africa have increased in recent years with about 24 percent estimated to live on less than the "standard" one US dollar a day (Faurès and Santini 2008). The majority of the poor live in rural areas and about 80 percent of them are directly dependent on agriculture for their livelihood (Faurès and Santini 2008). To reduce and/or reverse the increasing trend of rural poverty and generate employment requires substantial investment in irrigated agriculture in Sub-Saharan Africa (Faurès and Santini 2008, de Fraiture et al. 2010). However, transforming Sub-Saharan Africa's agriculture also implies intervention in water control (e.g. building storage reservoirs) as lack of access to reliable water supply is one of the major limitations to crop production.

Third, coupled with population growth is the increasing global energy demand. Africa's hydropower potential is estimated at 1,750 terawatt-hours but is largely untapped (McCornick et al. 2008). However, many of the developed hydropower stations are located downstream of agricultural areas. Closely linked to the need for African hydropower development are the rising global food and energy prices which have attracted foreign investment in agricultural land in Sub-Saharan Africa. In addition to the much debated issue of land grabbing, foreign direct investment in agriculture in Sub-Saharan Africa is likely to increase agricultural water use and this could lead to further enhancement of an already stressed water situation (Berndes 2002). Increase in energy demands does have implications on water allocation among competing sectors and this is a challenge to river basin management (de Fraiture et al. 2010).

Fourth, climate change is likely to compound the above challenges. Changes in temperatures, shifting patterns of precipitation, and changes in the frequency and intensity of extreme events will impact on water availability in a catchment (de Fraiture et al. 2010). In semi-arid areas of Sub-Saharan Africa, the use of water storage infrastructures and conservation technologies may make the difference between a saved crop and total crop failure. Empirical evidence shows that uptake of such storage and resource conserving technologies can bring about sustainable agriculture for local communities where rainfall is inadequate and water shortages limit crop production (Bossio et al. 2011). However, these technological innovations have direct impact on the rain water partitioning at the plot scale and lead to increase water use and competition. Downstream farmers may experience a decrease in water flows as a result of adoption upstream. So much as these technologies enhance rural livelihoods they may also increase the asymmetrical interdependencies among users in a catchment. It is thus vital that water management institutions take into account the differential availability of water, green and blue, among the variety of users at all scales.

1.2 CONCEPTS AND THEORIES

1.2.1 Responses to water management challenges

Many concepts and theories have been recommended to overcome the challenges of water resources management highlighted in section 1.1. These concepts and theories include Integrated Water Resources Management (IWRM), river basin management, institutional design principles, institutional bricolage, and property rights.

IWRM is generally accepted as a framework and approach to realising equitable and sustainable water resources management. Through Integrated Water Resources Management a more regulatory approach to water governance is being tried by many governments (GWP 2000). In this approach, the ownership of water is vested in the state and various forums and levels for stakeholders' participation in decision making related to water use are being provided. According to Bolding, (2004) global discourses on IWRM have endorsed three shifts in water governance: 1) less state but more market-driven regulation; 2) delegation of functions to the lowest appropriate level and involving stakeholders in decision-making; and 3) from administrative to basin management (management along hydrological boundaries) (Bolding 2004). An effective coordinated management of the water resources of a river basin depends on the presence of an institution whose regulatory mandate and tasks are known and accepted by a majority of stakeholders. Stakeholders can then be considered those who have a legitimate claim to the water resources.

As a natural unit, river basins are seen as the logical unit of water management and the space in which IWRM approaches can be realised. River basin management is designed to address the effects of upstream and downstream interdependencies of water use in a catchment (Moss 2004). The challenge is that a river basin is not just a simple spatial entity but a complex one. In terms of space, most river basins comprise several smaller catchments ranging from the scale of transboundary, sub-national or regional to local scale, nested within one another, each presenting unique water management problems and affecting the choice and functioning of water management structures (Bohensky and Lynam 2005). Replacing existing institutional units by institutions oriented around biophysical systems have been criticised as only leading to new boundary problems and fresh mismatches which raises the problem of institutional interplay (Moss 2004). According to Moss (2004) institutional interplay refers to boundaries problems related to political responsibilities and social sphere of influence, and that it is along these boundaries, where the jurisdictions and interests of actors overlap, that conflicts between institutions arise. Similarly, Warner et al. (2008) argues that water management approaches are not cast in stone but outcomes of political choices which is based on values and preferences. The choice of a river basin as the most appropriate scale of water management is just a political one, it can be made differently (Warner et al. 2008).

Governments presently are focused on introducing decentralised decision-making bodies such as River Basin Authorities, with prescriptions for public and private sectors involvement in decision making. Central to the approaches are some key design principles contending that management institutions can be crafted by the resource users and policy makers (Ostrom 1990, Ostrom 1993). Through analysis of self-governing institutions, Ostrom (1990) identified eight general design principles by which collective action institutions can be crafted. Crafting is considered a process of developing optimal institutions. The design principles have so far been used extensively in the water sector reforms in developing countries. However, resource variability and user mobility common in semi-arid regions highlight limitation of this approach and as result it does not necessarily translate into effective institutions and sustainable use of water resources (Cleaver 2000, Quinn et al. 2007). Based on the observed shortfalls, researchers have concluded that institutions elude design, citing that institutions may operate intermittently, in an ad hoc fashion through informal relationships, but may still be enduring and approximately effective (Cleaver and Franks 2005, Cleaver and Toner 2006). Actors involved are likely to rework the new arrangements in combination with the pre-existing local institutions, or reject them all together, a process which has been analysed as institutional bricolage (Cleaver 2002).

In questioning the validity of the design principles, scholars have conceptualised institutions as a dynamic product of social and political practices; as sites where authority is contested and negotiated; or as part of the interplay of knowledge and power (Mehta et al. 1999). The concept of institutional bricolage is increasingly being used to understand the transformation of diverse forms of social institutions. Although institutional bricolage sensitises the need for new institutions to be sufficiently embedded in existing local practices, this does not guarantee it will lead to equitable access and sustainable water management.

However, when water is scarce allocation arrangements emerge; such arrangements imply that certain claims to water are recognised by other users of the same resource. This recognition of one's claims to water by others as legitimate forms the basis of water property rights. According to Bromley (1991), a property right is the capacity to call upon the collective to stand behind one's claim to a benefit stream. Property right constitutes a bundle of rights which include: access, withdrawal, management, exclusion and alienation rights (Bromley 1989). But with these rights also comes the duty to respect similar rights of others (Potkanski and Adams 1998). The challenge however, is that sources of water rights are diverse and often conflicting. A river basin's water resources may be subject to regulation by more than one legal system. The rules and norms mediating water access and control may arise from local customs, donor projects, religions, and/or may be sanctioned or introduced by the state. There is thus a likelihood that these varied forms of water right claims co-exist and interact through a process termed legal pluralism (Meinzen-Dick and Pradhan 2002). The diversity of water rights may lead to struggles over whose right or claim is considered the most legitimate.

Overall there is no doubt that solving competition and conflict requires governance arrangements that can ensure equitable and sustainable use of the limited water resources. These include rules of how water is shared among competing users and the institutional arrangements to monitor and ensure compliance with the allocation mechanism. The governance process can be undertaken by the government, resource users as well as by organisations of all types and at all scales (Blomquist 2009). The challenge however, is that no perfect governance arrangements to be applied in water stressed river basins exist. This is even more challenging in semi-arid areas whereby a growing human population, variable water resources distribution and its finite nature result in potential conflicts between resource users. As a consequence many semi-arid countries are searching for appropriate management models that would provide a conducive environment for equitable and sustainable water resources management.

In Sub-Sahara Africa, countries have reviewed and/or redesigned their water institutions deriving insights from developed nations' theories and experiences with river basin management. Most governments attempt to formalise the water allocation system and river basin management arrangements. Users are obliged to apply for a water use right and to pay an annual user fee to a designated basin water authority.

The water right grants the right to use a certain amount of water, at a particular location and duration. Issuing water rights is claimed to lead to orderly use and proper care of the resource (Challen 2000). The governments' water law also grants priority rights to certain uses or users. The main rationale is that under scarcity allocation should be to uses with the highest returns per unit of water (basic human needs and environmental flows inclusive). In addition, participation by water users in the decision making process is considered to improve the administration of water rights and the management of water conflicts. Scholars have called for the creation of platforms or arenas (e.g. catchment forums and water user associations) through which diverse users can dialogue over water. However, the outcomes of the water reforms are far from the ideals being prescribed in policies. A disconnect between the local resource management reality and government-led water management institutions appears to exist.

This disconnect may be because government-led water policies and institutional arrangements fail to take local water management practices into account. Some local water management practices are well known historically, especially in (semi)arid regions as these developed into successful institutions for sharing water (see: Gray 1963, Fleuret 1985, Wade 1988, Ostrom 1990, Grove 1993, Adams et al. 1994, Ostrom 1998, Potkanski and Adams 1998). Local water institutions seem to offer flexible solutions to the problem of variable water supply arising due to climatic and weather changes (e.g. drought). Such locally evolved governance approaches, if well understood, could potentially be a substitute for, or used to improve, catchment institutions for water resources management being implemented by many governments.

Adopting this approach implies that local level approaches would be up-scaled while state-led institutions be down-scaled. Local level institutions would form the foundation and building blocks of catchment and basin-wide institutions. However, up-scaling local institutional arrangements requires understanding why they emerge, and how they function and are being sustained and the scales at which they remain effective.

This thesis wishes to contribute to this project by studying one river basin, namely the Pangani river basin, Tanzania. The Pangani basin is a partially closed river basin, partially because some of its tributaries do not flow throughout the year due to over use. It is partially open in that groundwater use is still limited there is very little knowledge on its use, availability, and interaction with surface water. It is a basin where state-led interventions date back from the colonial era and local practices evolved over a period of more than 100 years. This thesis focuses on the emergence and evolution of these institutions.

This thesis uses the terms "local" and "state-led", instead of frequently used terms such as formal and informal, foreign and indigenous, modern and traditional. The latter terms are often sources of contention when one attempts to distinguish between water institutions. In this thesis the term "local" is used to refer to farmer-initiated (endogenous) water management practices that evolved over time, and "state-led" to mean government-initiated forms of water management which is normally backed by statutory law and formal policy. By using local and state-led I avoid the use of even more problematic binary categories such as informal/formal or traditional/modern.

1.2.2   Upscaling local self-governing water institutions

Many scholars now argue that effective water institutions may be achieved by up-scaling nested arrangements in which local communities have been managing their water resources from homestead, plot, village, and sub-catchment levels (Van der Zaag 2007). Local communities do recognise their interdependencies and in return adopt and discard rules including management strategies as and when they require, by integrating history, and cultural meanings into management of water and conflict (Fleuret 1985, Potkanski and Adams 1998, Mohamed-Katerere and Van der Zaag 2003). Despite the exemplary theoretical and empirical base, lacunae continue to exist. For instance it is not clear why and how local cooperative institutions emerge and survive in the long run (Agrawal 2001). There is a puzzle on the relationship between collective action and inequality, heterogeneity, interests and power (Baland and Platteau 1999, Varughese and Ostrom 2001, Naidu 2009). It is important to understand how the water institutions form relational effects that can be successful in certain contexts, and fail in others.

Recently some scholars have attempted to explain why cooperation emerges in situation of water asymmetry. According to Van der Zaag (2007), the recognition by actors in a more advantageous position of their dependence on the cooperation of those in a less advantageous position motivates the former's willingness to forego immediate and short term benefits in order to secure long term benefits; this has been called 'hydrosolidarity' (cf. Falkenmark and Folke 2002). Theoretically, Van der Zaag's position can be explained from the theory of New Institutional Economics (NIE) and Assurance Problem (AP). NIE posits that individuals will carefully and rationally evaluate expected costs and benefits of their actions and cooperate when the expected benefits outweigh the transaction costs of not cooperating (Ensminger 1990). From the AP perspective, if enough people (critical mass) in a village are assured that others will cooperate with respect to a resource use, then incentives for the individual to respect them and also cooperate are high (Runge 1986). This is said to be true for communities that critically depend on a local resource base and face a high uncertainty with respect to that resource. Because of the uncertainty, they are more likely to develop collective arrangement as this may become cost effective for the whole group; can be efficient in allowing temporary access to other resources; and may also act as a safety-net for the community (Fleuret 1985, Runge 1986, Potkanski and Adams 1998). In this respect, users would forego immediate and short-term benefits even if the relative benefits of cooperation accruing to individual members of the group on average are somewhat less than under a system of exclusive use right (Runge 1986).

However, since the costs of collective actions are usually incurred in the current period while the benefits may only come later individual willingness to cooperate is likely to be affected by the present value of the benefits. Baland & Platteau (1999) report that the present value of the net expected benefits depends on the structure of users' time preferences. It is important to know at what discount rates and time horizons individuals are still prepared to forego immediate and short term benefits in order to secure long term benefits. But it is possible that users' recognition of their interdependence shapes their time preferences. For instance, actors likely to have a long history of interactions may value future benefits more than present options, so they continue to contribute to collective activities. However, actors that are not sure of continuous interactions, may heavily discount the future and may not cooperate. In addition there is also the issue of poverty that may mediate the users time preferences. The poor may not be able to afford to wait for later payoffs. Users' recognition of their interdependencies can be hypothesized to increase the chance of local water management practices to emerge and endure.

Also implied in Van der Zaag's (2007) definition is the spatial dimension of local water management practices, in that upstream user have to act in solidarity with their downstream counterparts. Over what spatial distance does solidarity still function? It is hypothesised that an increase in spatial distance between the users decreases the number of users who are assured that others will cooperate with respect to the resource use. Thus distant users are less likely to develop or sustain solidarity based water institutions. This may be explained by the fact that cooperation may be easily achieved in small-scale societies dealing face-to-face with well-known individuals, often even kin-folk, and by repeated dealings, but this natural type of exchange may vary, and diminish in intensity, with spatial distance between actors (Ensminger 1990). Are local water management practices then applicable to situations where distant villages share a water resource? From the previous paragraphs, it appears that to capture gains from local water management practices at larger spatial scales, it may be necessary to develop more complex institutions which ensure that people, who have no previous knowledge of one another, no kin relations, and perhaps no prospect of future dealings, will cooperate in good faith (Ensminger 1990).

It is possible that the solution to the complex problems in a catchment partly depends on the users' recognition of their interdependencies on one another. This requires one to look beyond the water flow. Hydrosolidarity may be a vital concept to explain the sustained existence of local water institutions. But there are still a number of questions on the success of local cooperative arrangements and the possibilities for their operation at different temporal and spatial scales.

1.3   RESEARCH OBJECTIVES

The overarching research objective is to explore conditions for reconciling state-led institutional arrangements and local water management practices. The above objective can be divided into the following sub-objectives (Figure 1.1):

First, attempt to understand the impacts of state intervention in catchment water management and its interaction with local water management norms and practices between neighbouring villages, between distant villages and within a catchment.

Second, attempt to understand local water management practices: why they emerge, and how they function and are being sustained. In other words attempt to understand the mechanisms that drive cooperation at the local level (e.g. turn taking in villages – between farmers sharing a furrow, between two furrows).

Third, develop a game theoretic model for considering alternative scenarios for collective catchment management.

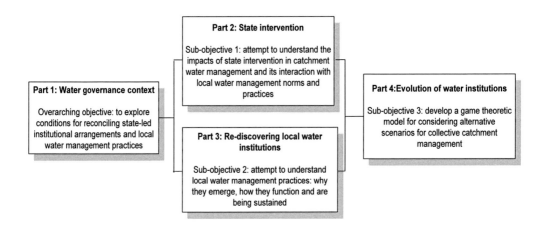

Figure 1.1: Schematic representation of the thesis research overarching objective and its linkage with the sub-objectives.

1.4 METHODOLOGY

"Go to the ant, thou sluggard; consider her ways, and be wise! It has no commander, no overseer or ruler, yet it stores its provisions in summer and gathers its food at harvest." Proverbs 6:6-7

Understanding why local water practices emerge, how they function and are being sustained requires following the water and the actors. By observing what the water users are doing and by asking why, hopefully one can learn. I used a case study research approach but not without modification. A case study strategy is considered suitable for empirical enquiry of a contemporary phenomenon within its real-life context, particularly when the investigator has little control over events (Yin 2003). My interest in adopting the case study methodology was to achieve analytical generalisation (i.e. use existing theories as a template to analyse empirical results) and not to achieve statistical generalisation (i.e. a small representative sample is used to make inference about a given population). By adopting this approach, I acknowledge the fact that any knowledge is partial and situated, i.e. a detached observer position is not possible to attain (Nightingale 2003). I attempt to emulate a reflexive model of science that embraces engagement with the subject of study (Burawoy 1998). Since the objective of the research is up-scaling local water management practices (i.e., moving from the micro to the macro level), the extended case method put forward by Burawoy (1998) was chosen. According to Burawoy (1998), "the extended case method applies reflexive science to participant observation to extract the general from the unique, to move from the micro to the macro, to connect the present to the past in anticipation of the future, all by building on pre-existing theory". An understanding of how and why local water management practices work at the micro-level (furrow) will first be established, drawing from existing theories and next an extension (scaling up) to the macro-level (catchment) will be attempted.

I studied the dynamics of water asymmetry, inequality and heterogeneity in a furrow irrigation systems; explored and analysed locally evolved water management practices between villages; attempted to understand larger scale operation of such water management practices for instance at the scale of a river; described the interaction between statutory and local rules. Moving further up, I followed the state-led process of creating water institutions at the catchment and sub-catchment scale; and attempted to explain how nesting of local practices within state-led institution are reinterpreted at various scales and levels.

1.4.1 Planned research approach

Initially the actor-network theory was the proposed theoretical framework for the research. Actor-network theory is a descriptive approach that seek to uncover "how" relations assemble or don't (Latour 2005). In describing and inferring from those relations actor-network theory develops a dynamic understanding of the relations. This approach was to be complemented by agent based modelling (ABM). The research was to be conducted through an iterative process. First an understanding of field level reality was to be established, data were to be collected about the selected catchments. This would start with an attempt to represent the dynamics of an observed furrow system in an agent based model. After comprehensive discussions with stakeholders and further field level data collection, a translation would be made to the level of arrangement between furrows in one village. This too would be complemented by comprehensive discussions with stakeholders in order to improve the agent-based model. Next the arrangement between actors in neighbouring villages sharing the water resources were to be constructed into an agent-based model informed by field data and discussions with communities. Finally, the whole catchment would be modelled and different water use sectors, stakeholders and other relevant parties consulted. In a recursive fashion different scenarios and sets of questions would be developed at each stage to improve understanding of local water management practices. To understand the state-led top-down approach, institutional mapping was proposed. This would involve detailed review of the legal, bureaucratic (organisation, their roles and responsibilities) and policy frameworks. Mapping of other state organisations, civil societies and transnational organisations involved in water management in the basin was envisaged to be done. This exercise would help highlight the interface between local level approach and government-led approach to water management. The outcomes of the agent based models would be confronted with the functioning of the large scale institutions being promoted by governments. The final output would be a synthesis of how water institutions may work in practice and the theories surrounding their effectiveness, sustainability or failure.

The above approach still seems valid to me, but reality was so complex that it eluded my research design.

1.4.2   Research methods and approach used

A recursive process that combines modelling and in-depth description of cases proved very challenging and requires a long time frame. Instead I have invested time in getting a thorough understanding of the water management institutions and practices through in-depth descriptions and analyses of selected cases within the Pangani river basin. The strategy used is inspired by the 'follow the water' approach (Latour 1988, Bolding 2004). The method can be summarised as a "safari downstream the Pangani river basin in search of actors" (cf. Asmal 2003). To identify the water users and their networks, I followed the water and in the process mapped all infrastructures and institutions around it.

Multiple case studies were conducted in the Pangani river basin, Tanzania[1]. The modularised approach is best summed up by using the metaphor of a big house with many semi-detached rooms, each of which can function independently but combine to form the complex whole. The case study sites are marked in
Figure 1.2.

I employed a mixture of techniques to achieve the research objectives and to be able to capture the dynamic interactions at the local level and produce narratives. I used qualitative methods such as participant observation to follow the day-to-day water management practices at the farmer level and observed the process of creating new institutions. I used in-depth interviews and focus group discussions to understand the practices (i.e. interviewed farmers, members of irrigation committees, NGOs and the basin management authority as well as local government officials). Although I interviewed many female water users, the majority of my respondents were male. This may have obscured the gender inequities and asymmetries in local water management. I used role play games and organised feedback workshops to engage in multiple dialogues with the subject under research and acquire more insights on the irrigation systems. I used quantitative methods such as cartography to link the

---

[1] A detailed description of the Pangani river basin, Tanzania, is provided in Chapter 2.

physical infrastructure and resources to the social relations that emerge around water use. Particularly, I mapped water infrastructures and agricultural land uses, using Geographical Position System (GPS), and conducted flow measurements. This allowed linking water sharing practices to social relations, such as kinship and trade. The research further benefited from in-depth analysis of grey literature on development project interventions collected from the archives of various organisations, department and the Pangani Basin Water Office. Finally, using game theory, I used the insight of local water management practices to develop a role playing game ("the irrigation canal cleaning game"). The game has been tested with master students at UNESCO-IHE and proved useful in understanding the dynamics of collective action in an irrigation system.

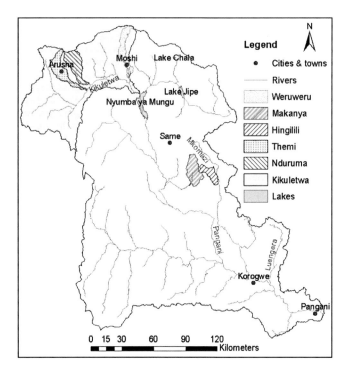

Figure 1.2: Selected case study catchments within the Pangani river basin, Tanzania.

1.5   SSI PROJECTS AND LINKAGES

This research is part of a larger research programme called the Smallholder System Innovations in Integrated Watershed Management (SSI) programme. Phase I was made up of six projects and the present research is project 6 (Figure 1.3). This research was conducted in close collaboration with the other five projects within the

SSI programme. Project 6 also provided the bridge between SSI phase I and II. Phase I was funded by the Netherlands Foundation for the Advancement of Tropical Research (WOTRO), the Swedish International Development and Cooperation Agency (SIDA), the Netherlands Directorate General for International Cooperation (DGIS), the International Water Management Institute (IWMI) and UNESCO-IHE Institute for Water Education. The SSI research was carried out by eight PhD, of whom six have completed their theses (Kongo 2008, Enfors 2009, Kosgei 2009, Mul 2009, Makurira 2010, Masuki 2011) and three Post-Doc (Enfors and Gordon 2007, Pachpute et al. 2009, Tumbo et al. 2011) researchers in two Southern Africa river basins, namely the Pangani in Tanzania and the Thukela in South Africa.

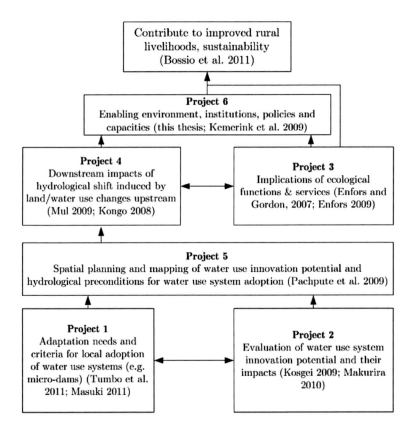

Figure 1.3: SSI research projects and linkages.

1.6 STRUCTURE OF THE THESIS

This thesis consists of four parts (Figure 1.4). Each of the parts begins with a short introduction. The first part (Part 1) comprises of two chapters (1-2). Chapter 1, "Introduction", sets the stage. It reviews relevance concepts and theories used in this research, provides the research objective and sub-objectives, and presents the research methodology. Chapter 2 (Komakech et al. 2011b) provides an introduction to the case study basin, "the Pangani river basin over time and space: on the interface of local and basin level responses". The chapter uses the concept of basin development trajectory to provide a general overview of the Pangani river basin and the major development challenges.

Part 2 of the thesis consists of three chapters (3-5). This section addresses research sub-objective 1 which focuses on state-led intervention in the water sector. Chapter 3 (Komakech et al. 2012c) deals with the impact of state-led formalization of water allocation systems on local water management practices. Chapter 4 (Komakech and Van der Zaag 2013) deals with water institutional design pitfalls. Specifically it discusses the concept of polycentric governance, institutional nesting, and catchment forums. Chapter 5 (Komakech et al. 2012e) deals with water transfers from agriculture to cities in the Pangani river basin, Tanzania.

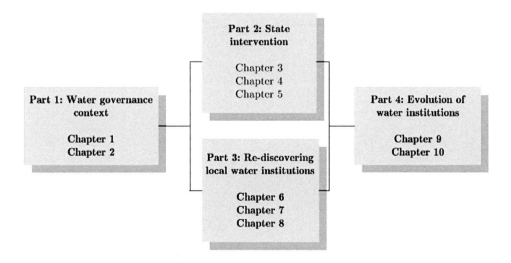

Figure 1.4: Structure of the thesis.

Part 3 of the thesis also comprises of three chapters (6-8). This section addresses the research sub-objective 2 which aims to understand local water management institutions. Chapter 6 (Komakech et al. 2012d) provides a detailed case study of the dynamic of water asymmetry, inequality and heterogeneity on collective action at the level of an irrigation canal. Chapter 7 (Komakech and Van der Zaag 2011) looks at

the emergence and functioning of local water management arrangements at larger
spatial scales. The chapter provides a case study on the emergence and functioning of
river committees in one sub-catchment of the Pangani river basin. Chapter 8
(Komakech et al. 2012a) presents a case study of the role of statutory and local rules
in allocating water between large-scale and small-scale users in one of the sub-
catchment of the Pangani river basin.

The last part (Part 4) of this thesis comprises of two chapters (9-10). Chapter 9
(Komakech et al. 2012b) addresses the research sub-objective 3 which concern the
development of a game theoretic model for considering alternative scenarios for
collective catchment management. Chapter 10 of this thesis provides the research
discussions and conclusions. This last chapter draws together all the findings from the
different parts of the thesis. In addition to addressing all the research objectives, I
highlight the contributions of this thesis to concepts, theories and research
methodology. The section also provides a critical reflection on the strengths and
limitations of the study.

Chapter 2

PANGANI RIVER BASIN OVER TIME AND SPACE: ON

THE INTERFACE OF LOCAL AND BASIN LEVEL

RESPONSES[2]

2.1 ABSTRACT

As the pressure on the water resources mounts within a river basin, institutional innovation may occur not as a result of a planned sequence of adjustments, but arising out of the interplay of several factors. By focusing on the basin trajectory this chapter illustrates the importance of understanding how local level institutional arrangements interfaces with national level policies and basin-wide institutions. We expand Molle's (2003) typology of basin actors responses by explicitly introducing a meso layer which depicts the interface where State-level and local-level initiatives and responses are played out; and focus on how this interaction finds expression in the creation and modification of hydraulic property rights. We subsequently apply this perspective to the case of Pangani River Basin in Tanzania.

The Pangani River Basin development trajectory did not follow a linear path and sequence of responses. Attempts by the state government to establish 'order' in the basin by issuing water rights, levying water fees and designing a new basin

---

[2] Based on on Komakech et al. 2011 *Agricultural Water Management* 98(11), 1740-1751

institutional setup have so far proven problematic, and instead generated 'noise' at the interface.

So far water resources development in the Pangani has primarily focused on blue water, and the chapter shows how investments in infrastructure to control blue water have shaped the relationship between water users, and between water user groups and the State. It remains unknown, however, what the implications will be of widespread investments in improved green water use throughout the basin – not only hydrologically for the availability of blue water, but also socially for the livelihoods of the basin population, and for the evolving relationships between green and blue water users, and between them and the State. The chapter concludes with a question: will green water development engender a similar double-edged material-symbolic dynamic as blue water development has.

The findings of this chapter demonstrate that the expanded typology of basin actors' responses helps to better understand the present situation. Such an improved understanding is useful in analysing current and proposed interventions.

2.2 INTRODUCTION

Integrated water resources management in the Pangani River Basin, Tanzania, is not a new phenomenon but a continuation of past attempts to control access and use of the basin water resources in time and space. In the pre-colonial era water resources management was governed by a customary system that was dynamic and aligned with the realities of the indigenous water resource users (see for instance Grove 1993, Adams et al. 1994, Håkansson 1998, Sheridan 2002, Tagseth 2008). German and British settlers introduced formal water law and in the early 1900s all water resources (ground and surface waters) were formally declared vested in the colonial government (Maganga 2003). This secured water rights for the settlers and allowed agricultural intensifications around the Kilimanjaro area (Van Koppen et al. 2004). The independent government of Tanzania (then Tanganyika) further upheld this principle, declaring all water to be vested in the United Republic under the Water Utilization (Control and Regulation) Act of 1974 (section 8) (Maganga 2003). The government introduced the concept of river basin management as early as 1981, nine basins were delineated and Basin Water Boards and Water Offices were created to manage the water resources (Maganga 2003). However, by 2009 the state-initiated water right system was not yet fully operational in the nine gazetted basins of the country. Since the colonial era, local water management approaches have continued to evolve, somewhat in parallel with, but separate from, the State-initiated institutional structures, creating a mismatch between the local reality and national policies.

The Tanzania National Water Policy of 2002 (Tanzania 2002a), now supported by a Water Act of 2009, attempts to bridge this gap (Tanzania 2009). This policy promotes active user participation in basin water resources management through formal representation in river basin water boards and by creation of catchment water committees and water users associations. The catchment committees are supposed to provide the linkage between local level user associations and the basin-wide water board. However, their establishment is yet to be fully operationalized. Only few Water User Associations (WUAs) have been established and only in some of the nine basins (Vavrus 2003, Mehari et al. 2009). Even then the WUAs creation processes are bogged down by unclear registration procedures and some of the registered users do not feel that the association is legitimate. Attempts to link up with local level institutions have so far proven problematic (Mehari et al. 2009). The majority of water users are poor smallholder farmers and livestock keepers who manage their self-initiated canal irrigation systems, locally known as furrow systems, without State support, and therefore do not feel the need to obtain State water rights and to pay water fees (Maganga 2003, Mehari et al. 2009). A common understanding of key concepts (e.g. state ownership, water right etc) and realities on the ground is generally lacking.

The Pangani River Basin provides an opportunity to test the typology of basin actors responses developed by Molle (2003) and modified by us in this chapter. According to Molle (2003), actors' adjustments (induced and intended) to water related problems do not necessarily follow a natural order or sequence. Rather a variety of strategies may be employed by individuals, groups and the State informed by context specific factors. The development of a river basin therefore depends on a multi-level response of society and the State to water problems (e.g. scarcity, floods and pollution), which shapes and is shaped by the nature of property relationship between the water actors (Molle 2003). We expand Molle's typology of responses by explicitly introducing a meso layer which depicts the interface where State-level and local-level initiatives and responses are played out, which finds expression in the creation and modification of hydraulic property rights. We also explicitly include an additional outer shell to account for important global and regional influences.

The chapter first reviews and expands the concept of river basin trajectory and focus more on the less well defined basin actors responses (section 2.3). It then provides an overview of the development of water use and institutions over time in the Pangani River Basin (section 2.4). The following section analyses the Pangani trajectory by focusing on local-level and State-led initiatives, and how these interact (section 2.4). The chapter concludes that the expanded typology of basin responses as developed in this chapter helps to better understand the present situation. It is suggested that such an improved understanding is useful in analysing current and proposed interventions, and the likely dynamics these will engender.

2.3   CONCEPTUAL FRAMEWORK: RIVER BASIN TRAJECTORY

River basin water resources are mobilised for domestic use, food production, energy and industrial development, and increasingly tourism. However, with a growing population and other intervening phenomena such as climate change, water resource availability changes over time and space. The responses of users and the State depend on their perception of the magnitude and seriousness of these challenges (Molle 2003). However, each time a decision is implemented, either locally or by the State, the river system is influenced and may change in terms of quantity, quality or timing of water flows, and other users are likely to be affected somewhere else in the basin. Over time the river system reaches "closure", a state of full utilization of available water resources, such that there is little or no margin for further development in one area of the basin without affecting user demands in another part (Keller et al. 1998, Falkenmark and Molden 2008). The development path of a river basin over time and space until closure is considered the basin trajectory. According to Molle (2003), the basin trajectory is a graphical representation that acknowledges the variety of micro/local and macro/global responses to water problem. Molle (2003) presents a conceptual model of the basin actors' responses (Figure 2.1). The outer circle represents the State domain, while the inner circle depicts the local level domain, all of which are geared towards the water related problems in the centre.

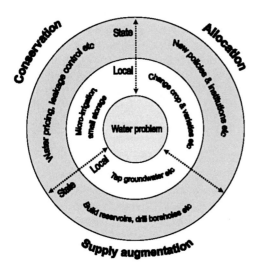

Figure 2.1: Types of actors' responses to water scarcity (Molle, 2003).

The framework breaks down the micro- and macro-level adjustments into three categories: supply responses; conservation responses; and allocation responses (see Figure 2.1 and Table 2.1). No *a priori* assumption is made about the possible sequencing of the responses as is often highlighted in other frameworks for river basin development (Keller et al. 1998, Molle 2003).

Table 2.1: Adjustments to water related problems in a river basin (Source: adapted from Molle, 2003).

Category	Description
Supply responses	Basin problem solved by augmenting supply from existing sources (building of storage reservoir) and tapping additional sources (e.g. inter-basin transfer).
	Locally, farmers may tap shallow aquifers; invest in local storage; and/or start conjunctive use of water from runoff, rivers, ponds etc.
Conservation responses	The key phrase is efficiency – using existing sources better without increasing water supply. Line agencies may implement: structural measures (canal lining, reuse of wastewater, leakage control etc); non-structural measures (improve dam or canal management); and establish rotation or other arrangement for better scheduling. The state may devise and enforce policies that elicit water savings (e.g. water pricing, rotation and quotas, supply innovation on plot level water management, improved varieties and cultivation techniques).
	Creation of innovative institutional arrangement to improve coordination between users (e.g. catchment forum, water users association, water boards).
	At a local level, users may: shift cropping calendars; adopt better cultivation techniques (e.g. mulching, shorten furrows etc); change crop varieties; invest in water saving technologies (micro-irrigation).
Allocation responses	Strategy could include reallocation of water between users of the same sector or across sectors either to raise water productivity or enhance land productivity, food security or equity, or reduce conflict and protests.
	At the local level, users may use the limited water for crops with higher return per m^3. At irrigation systems, allocation may be based on a set of factors (e.g. water duty per hectare, location, soil type etc).
	At the basin level, water may be reallocated to a given priority system, the rationale often being economic value. Molle (2003) notes that such allocation decisions are not only sources of potential conflicts but sometimes mediated by markets or negotiations. User groups within an area (e.g. catchment or irrigation system) may renegotiate rights to ease tension.

In Table 1, the three categories of responses are summarised and State and local level responses are distinguished. Measures taken at these two levels may not be complementary (dotted arrows in Figure 2.1). Molle (2003) states that mismatches between micro- and macro-processes should not be treated as noise [3]. We argue that understanding the "noise" is a prerequisite for developing interventions that can lead to an improved and better coordinated water resources management for basins such as the Pangani. The framework shows how micro/macro-level responses are shaped by a number of driving forces and the importance of State/citizen relationships, which defines the scope and room for manoeuvre and adjustment allowed to the different actors in the system. However, State roles do change over time and may also lead to the introduction of new institutions and forms of organisation, which may lead to the capture of small communal systems by new actors (e.g. local lords, village chairpersons, large estates, etc taking over the management of locally initiated furrows). It is important to note that decisions taken

---

[3] The term 'noise' is used here to symbolize the uncoordinated, ad hoc, contentious nature of water politics through time and space. It may be considered an alternative for what A.O. Hirschman termed 'voice', i.e. political engagement (Hirschman 1970).

by these actors are not necessarily informed by monetary consideration alone but also by the other benefits and increased power that will accrue to the different categories of actors within, and sometimes beyond, the basin. For instance investments in large-scale hydraulic infrastructures (e.g. a dam for electricity generation) may benefit the residents of large cities outside the basin.

Although Molle (2003) acknowledges the importance of State/citizen relationship and power structures among others, his problematization does not include the dynamic linkage between actors' (micro/macro) response and property relations. The actors respond to water related problems by making adjustments to the water resources system. These adjustments could be in the form of investments to create new or modify existing hydraulic infrastructures (e.g. reservoirs, irrigation furrows, etc) that would abstract, store, and/or convey more water for certain use functions (Coward 1986b, Van Koppen et al. 2008). The actors' investments (e.g. through labour and capital) in the physical infrastructure mediate their individual or group claims to the water conveyed (e.g. in terms of its quantity, quality, site and timing). The process of establishing claims is termed "hydraulic property right creation" (Coward 1986b, Van Koppen et al. 2008). The point is that the creation and modification of hydraulic infrastructure also affects the property relations among actors, which become the social basis for collective action in performing various resource management tasks (Coward 1986a). Property relations include the different networks of rules, principles and procedures that structure control over and/or right of access to a resource. It also includes the conditions for allocation, exchange, sale, inheritance, as well as the exclusion of those who have not contributed. Thus, similar to Coward's (1986b) vision of irrigation development (Box 2.1), actors' response to water related problems involve a property rights creation process.

Box 2.1: Coward's (1986b) vision on the property factor in irrigation management.

According to Coward (1986b), irrigation development can be considered a process of property creation. As new objects of property (such as weirs, canals, and water rights) are created, new relationships among people engage with these objects of property emerge. Ownership of and responsibility for irrigation works (their operation and maintenance) nearly always coincide.

Locally initiated irrigation systems are often characterised by clearly defined ownership patterns. In situations where the government builds a new irrigation scheme or rehabilitates an existing one, the irrigators are normally excluded from the investment process. Perhaps the most fundamental consequence is that cultivators, as non-owners of the hydraulic property, are alienated from that property and may not act as though they are responsible for it (even though government may wish them to do so). If State investment occurs in settings with existing community irrigation facilities, the usual consequence is the disturbance or even destruction of existing property relationships.

Since the property factor is important in mobilising local action for irrigation activities, it is an important policy variable to be used in designing future irrigation development activities. The irrigation investment ought to proceed in such a way as to create hydraulic property for the group that is intended to operate and maintain the irrigation facilities that are created or improved.

Similarly, the State may not necessarily invest in new hydraulic infrastructure but adjust by introducing new water laws and/or policy. The State, for instance, can declare full ownership of the water resources, create new institutions for water management, introduce obligatory registration of all water uses and start issuing water licences, permit or rights. Such adjustments can also affect or change existing property relations between actors. The concept of State ownership and "modern" water rights themselves may become major points of conflict, more especially in basins where the local actors' claims to water are tied to prior investments in hydraulic infrastructure.

Property rights define entitlements to use a resource, in a particular place, time and quantity and also the responsibility for its sustenance over time (Meinzen-Dick and Pradhan 2002). As water is fugitive, it is difficult to have a fixed property right on it in one place. Like many objects (land inclusive), water is often subject to regulation by more than one legal system. It is true that any social field (village, groups, association, community, State) may generate or enforce their own separate rules and norms (Meinzen-Dick and Pradhan 2002). It is also possible for these various kinds of rules and norms to coexist and interact within the same social field (e.g. statutory, religious, customary, donor projects, organisation, and local norms). Such coexistence and interaction of multiple legal orders is termed legal pluralism (see Meinzen-Dick and Pradhan 2002, Maganga 2003, Maganga et al. 2004). Thus individuals or actors, depending on the appropriateness, local knowledge, perceived contexts and power relation, can make use of more than one set of rules to rationalise and legitimise their decisions or behaviour. To capture the link between property right and the different responses at the micro- and macro-level, we add property relation within Molle's framework, intersecting these two levels (Figure 2.2). The link between property claims and actor responses is dynamic and two-way: individual actors make choices about which laws to use based on the specific water challenges they face; and their choices affect, and perhaps create new, relations with other actors.

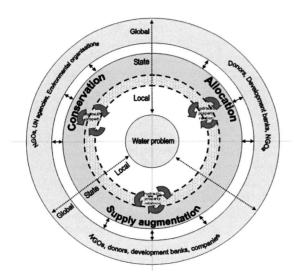

Figure 2.2: Modified actors responses and property relation (adapted from Molle, 2003).

In Figure 2.2, the curved arrows indicate that the property relation is dynamic both between the local actors and/or between local actors and the State. In addition, the cyclic nature of the arrow indicates that the processes of hydraulic property right creation can be both endogenous (local) and exogenous (triggered by government, non-local NGOs or other outside actors). The cycle is however imperfect and counter tendencies may frequently arise.

What is also not explicitly presented in Molle's (2003) conceptualisation is the global dimension impinging on the State thereby influencing the adjustments made at the micro and macro level. River basin actors do span beyond the hydrological and political boundaries and often link up with global players (e.g. international NGOs, donors, development banks, agencies, and markets). Some of these global actors make "invisible" adjustments to particular basin water problems. For instance, a donor country may support and finance responses that favour the creation of markets for its products in the receiving country. Lastly, it is important to note that lumping the adjustment into micro – macro level ignores the meso-level where the "middle game"[4] is often played. The meso-level lies at the interface where State responses often dynamically interact with the local level ones. The term interface is used here to refer to the point at which different and often conflicting life worlds or social fields intersect (Long 1999, Long 2001). The interface between the State and local level is framed through roles accorded to districts, wards, villages and locally evolved representation structures.

---

[4] Middle game is a chess analogy to refer to the most difficult phase of the game where a player needs to be well rounded to succeed.

2.4   WATER USE DEVELOPMENT IN THE PANGANI RIVER BASIN

2.4.1   Introduction to the Pangani River Basin

The Pangani River Basin is defined on the basis of hydrological boundaries, while the areas of jurisdiction of the Pangani Basin Water Board/Office (PBWO) include three other smaller adjacent basins, namely Msangazi, Umba and Zigi-Mkulumuzi. This thesis focuses specifically on the Pangani River Basin, and excludes the three smaller basins under PBWO jurisdiction. The Pangani River Basin covers approximately 43,650 square kilometres, 5% of which is in Kenya (Figure 2.3). The headwaters of the basin are located on the slopes of Mount Kilimanjaro and Mount Meru, with Kikuletwa and Ruvu rivers being the major tributaries of the Pangani river. The Pangani river passes through arid Maasai steppe, draining the Pare and Usambara mountain ranges (Mkomazi and Luengera tributaries, respectively) before reaching the estuary and the Indian Ocean. The basin covers all or part of four administrative regions of Arusha, Kilimanjaro, Manyara and Tanga. In total 14 districts and two major municipalities (Arusha and Moshi) rely on the water resources of the basin.

The current population in the basin is estimated at 3.7 million (IUCN Eastern and Southern Africa Programme 2009). The basin's population on the Tanzania side is influenced by in-migration of people in search of land and business opportunities (Mbonile 2005). The basin's population growth rate is about 4 % per annum, which is influenced by in-migration of people in search of land and business opportunities. The population of Arusha town, for example, doubled between 1988 and 2002, partly because of a booming tourism industry (Turpie et al. 2003, Mbonile 2005). About 80% of the population depends directly or indirectly on agriculture for their livelihoods. Local farmers in the highland areas (e.g. Kilimanjaro and Meru highlands) have practised canal irrigation for more than 200 years, and currently there are over 3400 known water users but the actual number is likely to be much higher (Mujwahuzi 2001). Intensive canal irrigation in the highlands compensate for the small farmland sizes (about 0.6 ha per household), while in the lowlands where agricultural land is abundant (about 10.4 ha per household) irrigation buffers against climate vagaries (IUCN Eastern and Southern Africa Programme 2009).

Figure 2.3: Map of the Pangani River Basin, with three neighbouring basins indicated (Source: Sadiki, 2008).

2.4.2   Water resources and present utilization

The climate of the basin is mostly related to topography. The flatter low-lying south-west half of the basin is semi-arid, while the mountain ranges have cooler and wetter conditions. The high altitude slopes above the forest line on Mt. Meru and Mt. Kilimanjaro receive more than 2,500 mm.a^{-1}. Southward rainfall varies from 650 mm.a^{-1} in the North Pare Mountains to 800 mm.a^{-1} in the western Usambara Mountains and 2,000 mm.a^{-1} in the eastern Usambara Mountains. On average the Pangani River Basin receives 34.77 x 10^9 m^3.a^{-1} of rainfall, of which about 55% is generated in the highlands of Mt. Kilimanjaro. The mean annual flow into Nyumba ya Mungu dam is approximately 43 m^3.s^{-1} (1.37 x 10^9 m^3.a^{-1}). Nyumba ya Mungu is one of the major dams (1.1 x 10^9 m^3 storage capacity, surface area of about 140 km^2) used for water flow regulation and generation of electricity in Tanzania (Turpie et al. 2003, Sadiki 2008). The outflow from Nyumba ya Mungu varies annually between 15–25 m^3.s^{-1} (Figure 2.4) and about 5–8 m^3.s^{-1} of the release is lost to evaporation, transpiration and consumptive uses between the dam and the hydropower station at Hale. According to PBWO (2007), the river discharge into the estuary is about 27 m^3.s^{-1} (0.85 x 10^9 m^3.a^{-1}) but other reports (Turpie et al. 2003, Andersson et al. 2006) put the figure between 10–18 m^3.s^{-1}. Evaporation, transpiration and consumptive uses account for the difference.

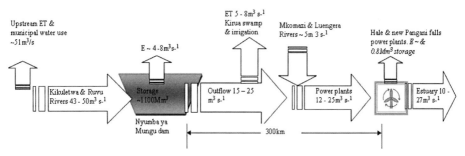

Figure 2.4: Water balance of the Pangani river basin (Adapted from: Turpie et al., 2003; Andersson et al., 2006; Beuster et al., 2006).

Groundwater potential is high in the volcanic and coastal aquifers, with boreholes yielding more than 100 $m^3.hr^{-1}$. Groundwater quality in the basin is generally good, with the exception of high fluoride content in some areas such as in the Arusha region. About 88% of the current groundwater abstraction in the basin is for irrigation purposes (Sadiki 2008).

The Pangani River Basin is considered water stressed with many of its tributaries only flowing for part of the year. Figures extracted from PBWO records indicate an estimated 71.7 $m^3.s^{-1}$ committed water of which 3.9 $m^3.s^{-1}$ is abstracted from groundwater. However, the figures are far from complete. Actual water use could even be higher. Water stress in the basin is attributed to: high population growth, increased cropping intensity, changing types of crops (e.g. paddy rice for sisal, flowers for coffee), low irrigation efficiencies (estimated to lie in the range 15–25%), uncoordinated development and possibly climate change (Turpie et al. 2003, PBWO/IUCN 2007, Sadiki 2008). Conveyance losses are considered a huge waste of water. A recent estimate of conveyance losses in Makanya catchment indicate that losses from the canals may range between 75% and 85% (Makurira et al. 2007). However, part of these high 'losses' feed the groundwater system currently being used by downstream actors in the form of spring water. The empirical understanding of the linkage between surface and groundwater hydrology is nevertheless currently wanting. Even at the scale of smaller catchments such as Makanya the interaction between groundwater and surface water system is not well understood (Mul et al. 2007b).

It is estimated that about 80% of Pangani basin population relies on agriculture, directly or indirectly, for their livelihood. Water demands are highest in the upper part of the basin, where it is used for agricultural, industrial, mining and domestic purposes, with 80% of the water abstracted in the Kilimanjaro region being for irrigation. Crops such as maize, rice, beans, bananas and vegetables are grown for domestic markets, while coffee, sugarcane, flowers, fruits and vegetables irrigated in large plantations or estates in the northern part of the basin are mainly for export. Water pollution is on the rise in the basin. The Rau tributary in Moshi, and Themi river in Arusha for example, are particularly polluted from raw sewerage and industrial discharges (Turpie et al. 2003).

Pangani River Basin has three installed hydropower plants at Nyumba ya Mungu (8 MW), Hale (21 MW) and New Pangani Falls (68 MW), contributing to about 17% of the national electricity demand (Turpie et al. 2003, Lein 2004). Although hydropower is largely a non-consumptive user, all of the Pangani river power plants are located downstream of major water users in the basin and are therefore potential sources of water allocation conflicts (Lein 2004, Mbonile 2005, Sarmett et al. 2005, Van Koppen et al. 2007). During water shortages, especially in the dry season, hydropower production can drop to as little as 30% of installed capacity (PBWO/IUCN 2007).

Conflicts between the different users and use sectors do occur. The Nyumba ya Mungu dam, constructed in 1966, has negatively affected wetlands. The largest wetland in the basin, Kirua, originally covered an area of 90,000 ha but this has reduced by about two-thirds (Turpie et al. 2004). At the estuary salt-water intrusion is increasing. Recent studies report decreased inflows, increased erosion upstream and flow modification by Nyumba ya Mungu and Kalimawe dams as the major causes of the increased salt intrusion (PBWO/IUCN 2007, Sotthewes 2008). Hydropower dams are holding sediments upstream limiting the amount of sediments reaching the estuary.

2.5   The Pangani trajectory: Local and State-led initiatives and their interplay

This section is organised in accordance with the proposed amended typology of responses: we start with describing local water development initiatives, followed by State-led water development initiatives, followed by a description of where and how these initiatives meet and interface. In so doing we test whether studying the actors' responses over time and space improves our understanding of present-day dynamics and policy dilemmas.

2.5.1   Locally initiated water management

The Pangani River Basin has an interesting history. Close to its source, the Chagga people have lived and farmed on the slopes of Mount Kilimanjaro for more than 400 years, while at the river estuary, Pangani was a cosmopolitan Swahili town several centuries ago, and a port linking Eastern Africa to the world through trade of coconut oil, millet, sorghum and slaves (PBWO/IUCN 2007, Tagseth 2008).

Since the 18th century or earlier, the Chagga people have constructed and maintained small-scale canal irrigation systems, known as furrow systems, that are considered the most extensive in Africa (Grove 1993, Gillingham 1999, Vavrus 2003, Lein 2004, PBWO/IUCN 2007, Tagseth 2008). Clan members established furrows and their founders automatically took charge of supervision of maintenance (organising users for repair works), arranging for meetings, allocation of water, and visually inspecting the furrows. For the management of the larger furrows, water committees and the founders (often chairing the committees) were responsible. Upon the death of a furrow's founder, a male descendant would inherit the position or was chosen by the furrow members as the new furrow leader (Gillingham 1999, Tagseth 2008). This type of leadership is still dominant in the highland areas, especially in the Kikuletwa catchment (Gillingham 1999, Pamoja 2006).

Although there have been significant attempts to control or even eliminate furrow systems, first by the colonial governments and then by the Tanzanian government, the furrow systems have persisted and instead spread to all other highland areas of the Pangani River Basin. Today over 2,000 furrow systems are reported to be active in the basin (Mujwahuzi 2001). Water rights in the furrow system are inheritable through the male line, based on one's labour and non-members can join through payment of small fees, often a local brew or in the form of sugar (Gillingham 1999, Sheridan 2002, Tagseth 2008). In a way one can argue that paying for water is not new, as even in the local system non-members paid and still pay entrance fees. However, it is a flexible system that can take many forms and charges are often affordable to the local users. Local water rights are based on one's labour; and involves rotational allocation, consensus building and conflict resolution, crucial features that appear to enhance and reinforce the legitimacy of these rights as well as their enactment (Maganga et al. 2004).

Water management in the furrow systems has continued to evolve, and in places where water scarcity is frequent, networks of relationships between different furrows sharing the same river have emerged. To date structures such as river committees can be found along many rivers in the Kikuletwa catchment (Pamoja 2006). River committees are self-initiated forums created by water users to allocate and monitor water use along a certain river reach during the dry season (see Chapter 7). River committees carry out water resources management functions and solve water related conflicts between furrows, commercial farmers, estates and villages sharing one river. They are informal in the sense that they are not identified as relevant structures in the new water policy – however, and interestingly, they are acknowledged by the Pangani Basin Water Office as important structures for improved water allocation and conflict resolution (Pamoja 2006). The committees are mainly active during the dry season when water allocation issues are prevalent. In cases where such a committee does not succeed to resolve problems, it often seeks help from government agencies (Komakech and Van der Zaag 2011). Elections of river committees are reported to be complex and representation issues do arise (Pamoja 2006, Komakech and Van der Zaag 2011). Campaigning for positions is often very competitive between the different water users and power plays may not be avoided. This may create tensions between the users (e.g. village versus estate; upstream versus downstream).

2.5.2   State-led water development – (a) infrastructure development

Besides the more than 2,000 furrow systems created by the local communities over centuries (Mujwahuzi 2001), the government has implemented a number of projects in the basin. First was the construction of a hydropower plant at Hale in 1936 by the colonial government. After its construction, Hale power plant was granted a water right of 66 years based on a minimum flow of 24.5 $m^3.s^{-1}$, which was later reviewed to 12.7 $m^3.s^{-1}$ on the basis of new insights from investigations (Lein 2004). The construction of Hale power plant put restrictions on upstream water developments. To overcome this constraint, Nyumba ya Mungu dam was constructed in 1966. The dam would allow for the development of irrigation (upstream and downstream of the dam) while securing minimum flows for the downstream power plant.

After food shortages as a result of, among others, the drought of 1974-1975, the government proposed to construct several irrigation projects to promote modern and efficient agricultural production. One was the Lower Moshi Irrigation Project (LMIP), which was constructed with assistance from Japan, and reached full scale operation in 1987 (Lein 2004). LMIP was designed to irrigate about 2,300 ha in four villages, of which 1,100 ha intended for rice cultivation, and would be granted a formal water right. The interesting point is that large parts of the command area had already been irrigated by means of 35 furrows developed by the villagers themselves. The project thus transformed a system of 35 furrow intakes into a unified irrigation system comprising two main canal intakes (Tarimo et al. 1998, Kissawike 2008). This set-up created various problems, the major problem being that the scheme only reached 30% of the farmers previously dependent on the same water. Those inside the scheme obtained state-sanctioned rights to irrigation water from the new hydraulic infrastructure. Many farmers who had previously created their own hydraulic infrastructure and were using the same water were left without such water rights. They nevertheless continued to use river water, in particular those located upstream of the scheme. Moreover, other farmers started to develop new irrigation furrows on their own, gradually leading to severe water shortages, first manifested in 1993. The problems were so great that the Lower Moshi scheme agency brought the matter to the Administrative Court, which, however, failed to provide a lasting solution since it could not ignore the customary rights to water that the furrow irrigators had acquired. The scheme agency subsequently started to look for alternative water sources. Attempts to augment water supply from Kikuletwa River met with stiff resistance from the Tanzania Electricity Supply Company (TANESCO), who claimed that water withdrawals from Kikuletwa River would seriously threaten the operation of its downstream hydropower plant (Lein 2004). The Pangani basin was closing.

In 1991 the Pangani Basin Water Board and Pangani Basin Water Office were established to register and regulate all forms of water use in the basin, implement the issuing of formal water rights to all existing users, and introduce and collect water fees from those users. It is often stated that the rationale for establishing the PBWO was to secure efficient and controlled water resources use but its timing, which coincided with the rehabilitation of Hale power plant, suggests that PBWO's mandate was linked to securing water for the downstream power plant that was of strategic importance for the Tanzanian economy 2004 (Fivas 1996, Lein 2004). In 1995 the government of Tanzania completed the rehabilitation of Hale hydropower plant and the construction of an underground power station at New Pangani Falls. The project was implemented despite the concerns that there would not be sufficient water for full operation of the new power plant. The location downstream of all major water users further restricts upstream development in the basin, especially for the irrigation sector. To overcome these challenges of water availability, the project was financed under the condition that the government would establish a functional river basin office responsible for overall water management in the basin (Fivas 1996, Lein 2004).

2.5.3   State-led water development – (b) water management

The German and British colonizers first introduced statutory water law in what is now mainland Tanzania in the early 20th century; all water was declared a state resource and the water right permit system was introduced (Grove 1993, Vavrus 2003, Sokile and Van Koppen 2004, Van Koppen et al. 2004, Van Koppen et al. 2007, Lein and Tagseth 2009). This formed the beginning of the State's intervention in water resources management in Tanzania. The evolution of state-led water management in Tanzania (summarized in Table 2.2) led to the introduction of water licenses, permits and rights issued by government authority to water users.

Although registration of water rights was first introduced by the colonial power to legitimize the settlers' claim for water, it was later used in the same way by large scale governmental and private irrigation firms, estates and parastatal organizations including TANESCO (Van Koppen et al. 2004). According to Van Koppen et al. (2007) the state-issued water rights were originally not meant for taxation, as it only stated the purpose of intended abstraction, the quantity allowed and the minimum quality of return flows. However besides providing information on water use, by imposing compulsory registration, any unregistered water use became illegal. In the 1990s volumetric assessment of water abstractions was subjectively done by the basin officers and the water rights system was not strictly implemented (Van Koppen et al. 2004). However, through subsidiary legislation (i.e. 1994 Water Utilization (Control and Regulation) Amendment Act No. 42 section 38 (2)), the government introduced a fixed once-off registration fee (Tsh. 40,000, about US$ 35) for all users and an annual economic water user fee in proportion to the volume of water allocated and type of use (a minimum of Tsh. 35,000 or US$31 for all abstractions less than 3.7 $l.s^{-1}$). Although it is often quoted that the introduction of the water fee system was to improve cost recovery for basin level management and to ensure 'wiser water use', the system was also meant to secure more water for the contested downstream hydropower plant (Fivas 1996, Lein 2004).

However, implementing the water rights and fee system in a basin where the majority of the water users are smallholder farmers who often already use and manage water under their own locally-developed water rights regimes is a significant challenge.

Whereas the relationship between taxation and wise water use may be straightforward for well metered systems such as in municipal water supply, for a river basin it is quite difficult and costly to implement. Blanket enforcement of water user fees can also result in people abstracting more water: in the Rufiji river basin people took payment for water as a license to use more of it - "we paid for water so we can use as much as we want" (Van Koppen et al. 2007). In such a context, limiting water fee payment to large scale users and to users who are able to derive significant economic benefits from the water use (e.g. hydropower companies, estates, water supply authorities, flower companies etc) could be more effective than a full implementation of the State's water right system covering all users (Van Koppen et al. 2007). The obligation to apply for a state water right for new proposed uses, however, may be an important management tool, especially for assessing the feasibility of the new use, whether existing uses may be negatively affected by it, and how this impact can be mitigated. Having noted all of the above, the main problem remains the limited capacity of the government to efficiently implement, enforce and monitor the proposed regularisation of water use.

Table 2.2: Historical development of formal water management in Tanzania (Source: Tanzania 2002a, Maganga 2003, Vavrus 2003, Lein 2004, Sokile and Van Koppen 2004, Van Koppen et al. 2004, PBWO/IUCN 2007, Van Koppen et al. 2007, Tanzania 2009).

Year	Highlights of formal law development
1890 - 1912	The Northern highlands area became of interest for European (mainly Germans, Italians and Greek) settlers who took about 200 km^2 of arable land and 567 km^2 of pasture land by 1913.
1923	First water law prepared and drafted during the German rule and adopted by the British. Water boards created. Water right was introduced to limit native water use and secure water for settlers.
1948 - 1959	Water control on paper became nation-wide. The colonial power declared absolute authority over water resources in the territory (Water ordinance chap. 257), and introduced (nominal) water right application fees (chap. 257 of 1948 Water Act section 35(d)).
	The 1948 Water Ordinance chap 257 and section 3 & 5 recognises earlier water rights including those under the 1923 Water Ordinance, lawful mining operations, some claims under the Indian Limitation Act and native law and customs. But only 'duly recognised representatives' of the natives were recognised.
	In the 1959 Water Ordinance the option of registration was extended to all users, including native water users who in earlier laws were recognised to have customary rights but this was given a second-class status.
	A centralised top-down government system for water management was maintained. The minister responsible for water appoints national water officers authorised to allocate and change water rights. The function was delegated to lower tiers of regional and basin water management institutions and water offices accountable upwards.
1974	The independent government declared all water in Tanganyika vested in the United Republic under the 1974 Water Utilization (Control and Regulation) Act, section 8. Created Regional Water Advisory Boards

Year	Highlights of formal law development
1981	The government amended the 1974 Water Utilization (Control and Regulation) Act and introduced river basin boundaries in place of regions as the first tier below the national level. The country was zoned into nine basins; each basin was under Basin Water Boards (replaced of the Regional Water Advisory Boards) and Water Office as the lowest tier.
1991 - 1993	Pangani basin water office was created in 1991 and the Rufiji basin office in 1993. The importance of the two basins for the nation's hydropower generation was the major driver for piloting river basin management.
1994-1997	Under influence of the World Bank, the amount for fees sharply increased. This was expected to promote the wise use of water upstream so that more water would be available downstream for hydropower.
1997	The Water Utilization (control and regulation) act was amended, giving mandates to Water Officers to inform and consider correspondence from Basin Water Boards before issuing water rights. Membership of the Central and Basin Water Boards to be drawn from the public, private sector, NGOs and women organisations.
2002	New National Water Policy devolved authority for water rights allocation to the catchment/sub-catchment level or even Water Users Association, although not yet operationalized.
2009	New Water Resources Management Act number 11 was enacted by the Parliament of United Republic of Tanzania. It provide for the institutional and legal framework for sustainable management and development of water resources; outlines water management principles; provide for prevention and control of pollution; provide for the participation of stakeholders and the general public in the implementation of the Water Policy. Importantly the new act gives more powers to Basin Water Boards. It redefined water right to water use permit, this was in respond to chaos the notion water rights induced in most basins in the country.

2.5.4   At the interface of the local and the State – (a) State-issued water rights and water fees in practice

The Pangani and Rufiji river basins were selected as pilots for the implementation of the Tanzania water policy of 2002. However, implementing the policy has not produced the much expected wiser water use and cost recovery. Since its inception, the PBWO carried out an inventory of water users in the basin. Presently the Pangani Basin Office maintains a database of over 3,400 registered and non-registered water users categorised as individuals, companies, village governments, institutions, and Water User Associations. Figure 2.5 gives the proportion of each user category. The term "user" applies to any person or group of persons abstracting water and it is used irrespective of the volume diverted.

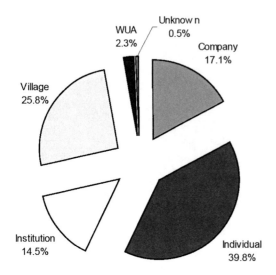

Figure 2.5: Pangani river basin water user categories (PBWO, 2008 personal comm.).

Table 2.3: Status and number of the different user categories in absolute terms (Source: PBWO database, 2008; personal comm.).

Status	Company	Individual	Institution	Village	WUA	Total
Application submitted	34	46	47	84	12	223
Water right granted Provisional	102	69	87	96	46	400
Water right granted Final	213	104	150	74	15	556
Without water right	70	996	127	628	5	1,826
Abandoned + Dormant	138	132	63	11	1	345
Superseded + Cancelled	18	23	8	5	0	54
Refused + Expired + Deferred + Withdrawn	18	14	17	0	0	49
Unknown status	3	3	5	0	0	11
Total	596	1,387	504	898	79	3,464

These figures are indicative, as many more users are yet to be identified. Table 2.3 shows that the majority of the individual farmers (996) and village governments (628) use water without State-issued water rights. Even WUAs, which are supposed to help with the water right administration, have mainly provisional water rights (46). It is interesting to note that a significant proportion of companies and institutions have been granted water rights. Possibly because of their size and profit motivation, these relatively large water users (e.g. large-scale commercial enterprises) have a greater ability to pay. Moreover, it is easier to identify and administer water rights for these companies and institutions. These figures therefore highlight the challenges of administering a water right system in a largely small-scale user environment.

The transaction costs in terms of personnel, logistics and means of enforcement outweigh the capacity of the basin office. Thus, although water user fees are planned to cover the costs of water right administration, it is currently inadequate to meet the operational costs of the Pangani basin water office and related activities (Figure 2.6).

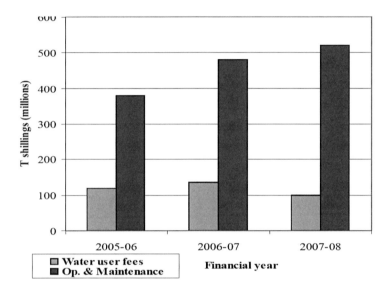

Figure 2.6: Revenue from collected water user fees and estimated operational (O&M) expenditure of PBWO (Source: Sadiki, 2008).

It can be noted that initially there was a steady growth in fee collection but in the period 2006 – 2008 there was a downward trend, an indication that a maximum collection level by the PBWO has been reached. It could also be argued that the reduction in fee collection is caused by users' realising that despite of paying for water they do not notice any improved access to water or a better service and see little reason to continue paying. The Basin Office, however, believes that by reviewing (meaning increasing the rates for some uses) the water tariff such that it depicts the economic value of water and by strengthening of WUAs sufficient revenue can be raised. The challenge is how to collect the increasing fees from the numerous reluctant small-scale water users in the basin. Since the transaction costs of collecting fees from large-scale users are comparatively low, Maganga et al. (2004) have suggested that the Basin Office prioritises these users. Such a strategy could show that it is possible to generate sufficient revenue to cover the cost of permit administration as well as all other operational costs of the basin office.

A final observation can be made about introducing water fees – what happens in the Pangani reflects a global trend. The premise seems to be that water fees not only automatically translate into more efficient water use, but also that basin organisations are able to, and should, finance themselves. It remains unclear how these normative ideas have permeated the policies and practices in the Pangani and in Tanzania as a whole, and why they are not frequently questioned.

2.5.5 At the interface of the local and the State – (b) Basin institutional setup

Water management in Tanzania is structured as shown in Figure 2.7. At the basin level the Basin Water Board is formally in charge of water management, with the Basin Office as its executive arm. At the catchment level, catchment committees are to be established, likewise sub-catchment committees should be established for sub-catchments while at the users level, Water User Associations are to be established to form the lowest level of state-led water management.

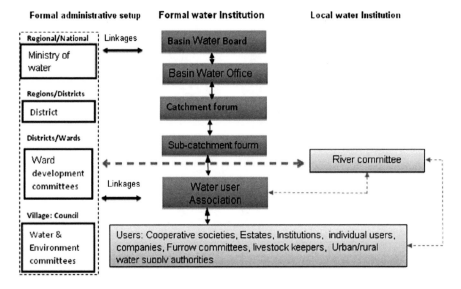

Figure 2.7: Basin institutional setup.

The organisational landscape in the Pangani River Basin is not exactly as formulated in the present water act (the Water Utilization Act of 1974 and subsequent amendments), nor as stated in the new 2002 water policy (Figure 2.7). Intertwined with the State-sanctioned setup (especially from the catchment to the community level) are the local arrangements, which continue to evolve. What is driving this dynamic are the investments in water infrastructure by a great variety of actors at various scales, for some to secure livelihoods through ensuring continued access to water (green and blue) for small-scale agricultural production, for others to exploit opportunities created by distant markets (e.g. horticulture), and others to ensure sufficient water and electricity for towns and cities. Currently there are diverse categories of water users and groups in the basin and a complex network of relationships exist between the various actors, who also interface with government structures over the management of water (Table 2.4). The description in Table 2.4 is far from complete as new forms of local arrangements are created as and when the users feel it is necessary to solve a critical local problem.

Table 2.4: Informal institutional arrangements in the Pangani river basin.

Name	Actors	Interests
River committee	Commercial farmers; estates; furrow committees; villages; urban water authorities using one river	Water allocation during shortage; conflict resolution; enforcement of allocation schedule; supervision of activities in the furrows; linking users and government agencies; and raising awareness. River committees could interface well with catchment committees and sub-catchment committees proposed in the policy.
Furrow Committee	Irrigators/domestic users; village governments	Water allocation and conflict resolution; supervision of activities; links users to river committees.
User group managing natural resources	Livestock keepers; NGOs; faith based organisation; other users	Secure access to resources; awareness; training on good practices.
Traditional leadership	entire community	Interface with daily life of communities. Customary law is the most important system guiding interactions related to water management. Traditional leaders are consulted during periods of water shortage. Perform conflict resolution task. Elders plays critical role in the conflict resolution
NGOs; Faith Based Organisations; CBO; Donors; Agencies	International and local organisations	Environmental conservation; water supply management; and tourisms. This group may create conflict with resource management if their interventions are not set within the context of integrated river basin management.

As discussed above, Pangani is a river basin where locally developed institutions interact with State-initiated organisations large-scale users with small-scale users; upstream users with downstream users and State actors deal with water users of various sizes and power. The Pangani River Basin is also a place where development partners, donors, international NGOs, environmental organisations experiment with alternative models of water management. Differences between these actors frequently exist or emerge, which sometimes escalate into conflicts. Conflicts may be categorised under three groupings – conflicts of ideology, conflicts of size and conflicts of location, which generally overlap (Table 2.5).

Table 2.5: Category of conflicts in Pangani basin (Adapted from: Mbonile, 2005; Sarmett et al., 2005).

Category	Description
Conflicts of ideology	Ideology considered here as a way society look at things. Local users in the basin are reluctant to apply and pay water rights, arguing that water is a "gift" from God. These communities reject government efforts to manage water resources and to charge for its use, even to the point of vandalizing water control gates and structures.
	In addition, juxtaposed against the local indigenous water ideology is the imported Integrated Water Resources Management (IWRM) ideology. In many areas local users have developed over time and space, water management approaches based on local culture and tradition. This age-old water ideology as found in northern Tanzania contrasts with the global IWRM ideology that emphasise the importance of cost-recovery, and full state control over water resources using imported measures and technologies. The mismatch tends to create conflict of what is the best approach to managing Pangani river basin's scarce water resource.
Conflict of size	Conflicts exist between users of different sizes & power in the basin. E.g. large scale irrigation plantations, often backed by foreign investment and using hundreds of litres of water per second through efficient drip irrigation system, differ starkly from small-scale users of traditional furrow systems with efficiencies as low as 14%. Similarly three urban centres in the basin require more water as they expand, pitting city municipalities against the village governments of farming communities.
Conflicts of location	Upstream – downstream conflicts occur. For instance, the Tanzania Electricity Supply Company (TANESCO) located downstream, pays a royalty to the Ministry of Water & Livestock Development for 95 MW, assuming a 45 m³/s flow. Because of reduced rainfall and upstream abstractions, the company often receives as little as 15 m³/s, limiting production to as low as 32 MW, forcing power-shortages that affect the entire nation, creating national-level tensions. Reduced flow also have led to saltwater intrusions in the estuary up to 20 km inland, compromising agricultural activities in the lower basin..

2.6 DISCUSSION

The Pangani River Basin trajectory does not follow a linear path and sequence of responses as often portrayed in a number of river basin development models (see for instance Keller et al. 1998). Instead, the path of basin development has been haphazard. Responses that parallel those applicable to water conservation were implemented at an early stage. The colonial government introduced the system of water rights not because it wasn't possible to carry out supply augmentation in 1923 but simply to legitimise their new claims to land and water resources in the basin and Tanzania at large. In fact the only large storage dam in the basin, Nyumba ya Mungu, was constructed in 1966 after a number of allocation and conservation responses had been tried in the basin.

Actors try out different forms of adjustments but it is only the responses that are successful and "stick", that are visible. Responses can be in the form of symbols, such as those laid down in legislation, or can be material, taking the shape of concrete river intakes, enlarged canals, lined water ponds and large dams. Responses often seem to combine symbolic and physical manifestations in specific ways. The actors devising those responses that have lasted must be considered the more influential. By being able to shape the environment to suit their needs they become powerful, often at the expense of those whose alternative responses did not "stick". This shows that problems of water scarcity not only trigger adjustments but also create new opportunities. As Swatuk quoting Allan (2003) puts it, water resources allocation and management is a political process and the outcome is always partial, because it results from the complex give and take of numerous actors negotiating in their own interests (Swatuk 2008). This is similar to the argument by Molle (2003) that decisions taken may not be solely informed by economic considerations but also by other types of benefits, such as increased power, that may accrue to the different categories of actors within society and sometimes beyond. The Tanzania Water Utilization Act of 1923 and subsequent amendments provide good examples as most of the conservation and allocation mechanisms initiated by the State were meant first to protect the interests of early European settlers, and later hydropower interests in both Rufiji and Pangani River Basins.

A second point to note is that actors' responses to water related problems affect the relationships among them. For instance, after Independence, the government of Tanzania (then Tanganyika) declared that all water resources within its territory belonged to the State, abolished management of water resources by clans, and introduced new water management structures among others. As a result, furrow systems constructed by clans became community property. This affected the relationships between the original clan and the new entrants, but not fully as intended: the roles of founding clans often remain important if not dominant (Mul et al. 2011).

A further point that the Pangani case reveals is that the impact of interventions of outsiders, such as the State, have local effects. As a basin is closing such impacts become larger, as well as the responses these may trigger from local actors. This is why we proposed to amend Molle's typology of basin actors' responses with a dynamic interface in-between the local and the State. Moreover, impacts and responses are played out in two different realms – symbolically and materially - simultaneously; whereby property claims are the symbolic expression of physical alterations of water and soils by humans. This is why we emphasised the importance of hydraulic property relations at that interface.

The amended typology of responses provided insight on the dynamics at the *meso-level* (i.e. interface of state and local level responses). The Pangani case shows how over time the government of Tanzania have attempted to create 'harmony' through new water policies, and laws that redefine roles of districts, wards, villages, and also creates basin water management structures (e.g. catchment committees, Sub-catchment committees, WUAs etc). The state initiated a water use registration process and obligatory water rights for all users. This has created large power differentials and contested claims. For example, large commercial estates, electricity company, and municipalities were able to acquire state-issued water rights more quickly than many local users (mostly furrows) who often remained without. Nevertheless, the acquisition of water rights did not immediately render the large users more powerful, as local users used different means to pursue their claims. For instance there have been cases in the basin were large scale investors (e.g. flower companies) are helped by a government agency – the Tanzania Investment Centre (TIC) - to acquire water rights but once the investors reached their new area, districts and local leaders often urged them to comply with the local system of water allocation . In addition, local politicians do get in the way of local users committees; for example, Ward Councillors were observed to interfere with the works of river committees along Themi and Nduruma rivers, some to the extent of disbanding the locally elected representatives. Thus instead of harmony, past adjustments have generated 'noise' at the interface. 'Noise' symbolizes the uncoordinated ad hoc contentious nature of water policies and laws through time and space, while 'harmony' symbolizes the Tanzanian government's attempts to restore order through rewritten water policies and laws. It is like a conductor trying to get an orchestra together and to play from the same sheet music. What Pangani case shows is that perhaps the sheet music itself is written for a certain part of the orchestra with the inevitable result that the rest of the orchestra is going to play its own favourite tunes thereby creating dissonance, i.e. noise (pers. comm. Larry Swatuk, 2009). Actions taken by the government since 1991 that were justified in terms of bringing order in the chaos have not succeeded.

Finally, sustainable allocation and management of the Pangani River Basin water resources is not and will not be an easy venture. Although the existing records indicate that the water resources are over-committed, they are far from conclusive and it is still not clear how much water is actually being consumed. Significant room for improvement in the rain-fed agriculture sector (green water use) exists, which is likely to increase the productivity of water and may enhance the livelihoods of many smallholder farmers. This will involve physical alterations of the landscape, such as terracing, other soil improvement measures through tillage, rainwater harvesting structures, requiring huge investments mainly in the land, likely having impacts on property relations. Such a shift towards increased use of green water will have impacts on downstream blue water availability. These impacts are however difficult to measure and predict, and the precise implications for other uses, including hydropower, are therefore unknown, although likely to be small or even insignificant (Ngigi 2003, Rockström et al. 2004, Kongo and Jewitt 2005).

2.7 CONCLUSIONS

As the pressure on the water resources mounts within a river basin, institutional innovation is triggered and evolves. This may not be a result of a planned sequence of adjustments, but an emergence arising out of the interplay of several factors. Given the complex institutional dynamics found in the Pangani River Basin, with clearly distinct bottom-up and top-down initiatives, the steadily increasing pressure on the basin's limited water resources causes friction at the interface. With increasing demands posed on the water resource local level arrangements inevitably will confront State-led initiatives and vice versa. This implies that there is a need for institutional arrangements at increasingly larger spatial scales that are able to incorporate, "commensurate" and fuse the differential values found at these different scales.

By focusing on the basin actors responses, this chapter has demonstrated the importance of understanding how local level institutional arrangement emerged, how they evolved over time and how they interface with national policies and basin-wide institutions that have been established more recently.

Drawing from the above discussion, the government of Tanzania faces a difficult decision over the allocation and management of Pangani River Basin water resources. The dilemma lies in whether priority should be given to local socio-economic development with a focus on livelihoods and food security (e.g. through investment in agriculture and green water use, improvement of furrows, small dams) or to economic development of the urban areas (e.g. by allocating more blue water to hydropower which happens to be located downstream of all other major water users).

This dilemma cannot be resolved on the basis of knowledge and information currently available. So far water resources development has primarily focused on blue water, and the chapter has shown how this has affected and shaped the relationship between water users, and between water user groups and the State. What remains unknown is the implication of widespread investments in improved green water use throughout the basin – not only hydrologically and for the availability of blue water, but also socially, for the livelihoods of the basin population, and for the evolving relationships between green water users, between green and blue water users, and between them and the State. A question is whether green water development engenders a similar double-edged material-symbolic dynamic as blue water development has.

PART 2:

STATE INTERVENTION: RECONFIGURING

PANGANI BASIN WATER INSTITUTIONS

Arusha Declaration monument
"State intervention not a Devine intervention"

There is a general believe that defining water use entitlements and crafting institutional arrangements to monitor its enforcement will achieve economic efficiency and social equity and will maintain or restore order in water stressed catchments. In addition it has been argued that the creation of catchment forums for users to dialogue and participate in the decision making related to water management will improve coordination and mitigate water conflict in a catchment.

Most governments have intervened in water stressed catchment by creating formal organization to register users, issue water use rights, levy annual user fees in a catchment. State intervention in regulating the use of scarce water resources, however, is not a Devine intervention. Such interventions do not necessarily achieve the goal of equitable and sustainable water use.

This section addresses the research sub-objective two: to understand the impacts of state intervention in catchment water management and its interaction with local water management norms and practices. The three chapters in this section describe and analyse the challenges and impacts of state intervention in the Pangani river basin, Tanzania. Using the concept of institutional bricolage, in Chapter 3 I discuss how newly created state-led institutions are likely to be re-interpreted and transformed at the local level. State formalization of property right as observed does not change the day-to-day water allocation rules significantly; instead it negatively impacts the existing local water management practices. An important finding in Chapter 3 is that the hydraulic position of users may be considered a driver for institutional innovation and change. It is always the disadvantaged downstream users that initiate the process of institutional change in times of water scarcity.

To overcome the challenges of institutional fit between state and locally created institutions the concept of nested polycentric governance has been proposed by scholars. Adopting a polycentric approach would allow larger catchments to be modularized into semi-autonomous governance subunits. However, this still raises the question of legitimacy and relevance of the created institutions to the local water users. In Chapter 4, I describe and analyse an attempt to restructure spaces for participation in the upper Pangani basin, a larger African catchment. Using the formation of the Kikuletwa catchment forum as a case, I investigate the relevance of the polycentric governance approach as a framework for integrating local-state institutions. A major finding is that polycentric governance does not work in this river catchment; at least not in the manner it is currently being implemented.

Case studies of city water capture in the Pangani river basin (Chapter 5) show that the government's priority allocation system provides opportunity for powerful users to dispossess other users of their water sources. The enforcement of water rights is challenged by flow variability and the difficulties of monitoring dispersed users rendering the right system ineffective.

Chapter 3

FORMALISATION OF WATER ALLOCATION SYSTEMS

AND IMPACTS ON LOCAL PRACTICES IN HINGILILI

SUB-CATCHMENT, TANZANIA[5]

3.1 ABSTRACT

This chapter analyzes the impact of one such government led formalisation process on local water allocation practices. Based on a field study in Hingilili sub-catchment, Tanzania, we find that government interventions do not achieve the goal of equitable and sustainable water management. However, we find that the principle of good neighbourhood that still exists between highland and lowland farmers in Hingilili could form a base to reconcile diverging water interests between highland and lowland farmers. The chapter shows that the concept of bricolage (Cleaver 2002) is useful to demonstrate the need for new institutions to be sufficiently embedded in existing local practices to succeed, but this is not a sufficient condition. The hydraulic position of the various actors (upstream or downstream) must also be taken into account, and may be considered a driver for institutional innovation.

---

[5] Based on Komakech et al. 2012. *International Journal of River Basin Management* 10(3), 213-227.

3.2   INTRODUCTION

The increasing competition over water observed in many river catchments in Sub-Saharan Africa puts additional demands on water institutions, and their capacity to reconcile competing claims. One response by governments is the formalisation of the property right to water. Formalisation includes the registration of water uses and users, the issuing of water rights or water permits to users, the levying of an annual water tax or water fee on permit holders, and the creation of formal organisations of water users sharing a common water source, often called water user associations (WUAs) (Van Koppen 2003).

The government of Tanzania was among the first countries in Africa to start the process of formalisation of water management through a comprehensive reform of its water sector. Formal policies, laws and institutional arrangements have been provided to achieve the objectives of equitable and sustainable use of the country's water resources. Key components of the reforms included decentralization of decision making to newly created Basin Water Boards, the formation of WUAs and the introduction of water pricing and cost sharing arrangements. As all water is vested in the United Republic of Tanzania, all water users are required to apply to the basin water boards for water use permits (Tanzania 2002a). Basin water boards and offices were created in the nine gazetted river basins in mainland Tanzania in 1989, and most of these boards have been issuing water use permits and collecting annual water fees from registered users for close to a decade.

There are mixed views on the performance of Tanzania's water policies and laws (Van Koppen et al. 2004, Lankford and Hepworth 2010). Although the water reforms may be considered an attempt to attain equitable and sustainable management, they are being implemented in an environment where local water management practices exist and have evolved over long periods of time. In all nine river basins farmers negotiate water-sharing arrangements at plot, village, and sub-catchment levels, drawing on local, customary and state-sanctioned practices (see for instance: Grove 1993, Gillingham 1999, Sokile and Van Koppen 2004, Van Koppen et al. 2007, Mul et al. 2011). In large parts of the nine basins local and customary arrangements continue to govern access to water, and even those users that have acquired state-issued water permits still are forced to draw on, and respect, local practices to gain access to water.

The Water Act of 2009 recognises customary water rights and management practices, which are considered to be of equal status to water use permits granted by government. However, customary users must still formally apply for water use permits with the basin water boards. Some scholars have argued that the current formalisation process duplicates management efforts leading to collisions of roles between state-introduced and locally evolved arrangements (Sokile et al. 2005). In addition, the introduction of water levies through the statutory water permit system has in certain instances exacerbated the competition and conflicts between different water users (Maganga et al. 2004). Lankford and Hepworth (2010) argue that the current approach is too regulatory and rigid, which does not adequately fit the conditions found in Tanzanian basins, such as high seasonality of water availability and demand, large spatial distances between users, and high evaporation losses.

In this chapter we explore and analyze the impact of the introduction of state-led water allocation and management arrangements in the Hingilili sub-catchment, Tanzania. We find that the new institutional arrangements do not achieve their goal of equitable and sustainable water use. The arrangements as implemented in Hingilili sub-catchment a) have not been able to solve existing water conflicts between the highland and lowland in the sub-catchment; b) have not achieved more equitable access to and control over water resources; c) have not been able to collect sufficient revenue and realise the principle of cost recovery; and d) are not yet well embedded in, and consistent with, pre-existing arrangements. Although the water users are now paying the annual water user fees, at the time of conducting (i.e. between November 2008 and December 2009) the research, they had not yet been issued with water permits.

Using the concept of institutional bricolage we show that the government initiatives and locally evolved arrangements are being re-interpreted and transformed at the local level. Actors interpret, utilize, contest and/or renegotiate the arrangements to fit with the local context; a process often termed socially embedding of institutions (Cleaver 2002). To influence this process an in-depth understanding of the local context is needed, as well as a careful, phased and tailored approach.

The remainder of the chapter is organized as follows: section 3.3 presents the concept of institutional bricolage as a theoretical framework. Section 3.4 introduces the case study sub-catchment. Section 3.5 presents the evolution of local water institutions in the Hingilili sub-catchment, starting from pre-colonial time through to 2003. Section 3.6 presents the most recent interventions in Hingilili since 2003. Section 3.7 discusses the research findings in light of the theoretical framework. Finally, section 3.8 provides some lessons on the impact of formalisation of water management on local institutions.

3.3 THEORETICAL FRAMEWORK: EVOLUTION OF WATER INSTITUTIONS THROUGH BRICOLAGE

As the available water in a river catchment becomes limited, clear rules are required that define who can access the resource, for what purpose, at what location, how much, when and for how long, as well as how these rules will be enforced. The process of defining these rules may be undertaken by governments, water users and nongovernmental organisations. Local users are able to develop self-governing systems that are robust and that lead to a sustainable resource system (Ostrom 1990). It is frequently argued that new institutions for resource management could be crafted in such a way that they dovetail with local practices (Boettke et al. 2008). One particular suggestion is to organise institutions in multiple layers of nested enterprises (Ostrom 1990). It is argued that nesting allows more inclusive organisation to emerge from the smaller, more exclusive self-organised units without the latter giving up their autonomy or being sidelined (Marshall 2008). According to Marshall (2008), nesting provides vertical assurance to lower-level agents and allows them to place greater trust in institutions of collective property rights that they self-organised. However, this reasoning seems to be based on the assumption that institutions being nested will not change in the process. First, nested units will have to surrender some of their autonomy to account for similar needs for recognition of other nested units. Second, actors involved are likely to rework the new arrangements in combination with the pre-existing local institutions, or reject them all together, a process which has been analysed as institutional bricolage (Cleaver 2002). The new institutional arrangements are thus likely to acquire new meanings, which may affect the robustness and self-sustaining character of the nested units. Formalisation thus requires careful institutional analysis.

The concept of institutional bricolage is based on the idea that institutions are constructed through analogies and styles of thought already part of existing institutions (Cleaver 2002). It recognises both the enabling and constraining aspect of institutions. It posits that individual action related to collective management is defined by both agency and structural constraints (Cleaver 2002). Institutional bricolage was first introduced by Mary Douglas in 1987, when she extended the concept of 'intellectual bricolage' to institutional thinking. She used it to argue that institutions are not necessarily the outcome of individual rational choice (Douglas 1986 as quoted in Cleaver 2002). Mary Douglas highlights the unconsciousness and messiness of institutional change, as opposed to deliberate and rational designing of institutions (Mehta et al. 1999, Cleaver and Franks 2005, Sehring 2009). Of importance to the process of bricolage is the aspect of overlapping social identities of the bricoleurs. Actors involved may call on a variety of attributes (e.g. economic wealth, special knowledge, official positions, kinship and marriage) to justify their institutional position or influence. Cleaver (2002) elaborates on two additional aspects of bricolage – 1) the practice of cultural borrowing and adaptation of institutions to multiple purposes; and 2) the prevalence of common social principles which foster cooperation (as well as conflict) between different groups of stakeholders. Institutional transformation is taken to be a complex and dynamic reconfiguration and recombination of existing institutional resources.

The concept of institutional bricolage is increasingly being used to understand the transformation of diverse forms of social institutions. Cleaver uses institutional bricolage to argue that the mechanisms for resource management and collective action are constructed from existing and new institutions, styles of thinking and sanctioned social relationships (Cleaver 2000, Cleaver 2002, Cleaver and Franks 2005). Although she sees actors as both conscious and unconscious social agents deeply embedded in their cultural context, she argues that they are still capable of analysing and acting on the problem at hand (Cleaver 2002). This is similar to Galvan (1997), who uses the concept of syncretism to reconcile the structure and agency aspects of both culture and institutions (Galvan 1997). Shering (2009) uses bricolage to analyze the transformation of water institutions in two post-Soviet states. She finds that actors involved influence the outcome of reform through selective adoption of certain rules they consider socially appropriate and neglect rules deemed incompatible with the existing logic. She shows that institutions emerging through bricolage may serve the interests of certain powerful actors thereby reinforcing existing social inequities. Earlier the concept was used to show how meaning often permeated from one context to another through sets of connected rules, a process termed "institutional leakage" (Douglas 1986 as quoted in Cleaver 2002). The use of existing rules, cross cultural borrowing or past ways of thinking in the construction of new rules also highlights the persistence of institutions over time and space - what has been term path dependency (see: Pierson 2000, Sehring 2009). Pierson (2000) argues that the benefits of staying along the same path increases over time and hence the relevance of considering historical events.

By using the concept of institutional bricolage we wish to understand how state-led water allocation and management is reinterpreted and transformed in Hingilili sub-catchment. The following section presents the case study sub-catchment and the research methods. It also sets the foundation for analysis of the transformation of the water institution over time in Hingilili sub-catchment.

3.4 RESEARCH METHODS AND CASE STUDY AREA

3.4.1 Methods

The objective of this research is to understand the impact of government led formalisation on the local water allocation practices in Hingilili sub-catchment, Tanzania. We explore how locally evolved institutions change over time and how these arrangements interface with state-led reforms. Fieldwork was conducted between November 2008 and December 2009 (in total four months of field work). The study focused only on smallholders' irrigation, the largest water consumer in the sub-catchment, but water is also used for domestic purposes, construction (brick making), and watering of animals.

We mapped all significant irrigation infrastructures in the sub-catchment and identified water committees managing water allocation. Through structured and semi-structured interviews with twelve canal committees (in total 96 committee leaders were interviewed), we collected historical information on the development of the irrigation canals, existing institutional arrangements, how water is currently allocated and conflicts resolved. We also made field observations of irrigation practices, and reviewed project reports and documents from nongovernmental organisation involved in irrigation improvement in the area.

We interviewed staff from Pangani Basin Water Office (PBWO), Same District council, Pamoja Trust (a local NGO) and the Traditional Irrigation Improvement

Programme (TIP) in Same district. The interviews at this level focused on the process of formalisation of water management in the Pangani basin and in the Hingilili sub-catchment in particular. We interviewed the leaders of the recently created organisation in Hingilili lowland (MUWAHI - Muungano wa Wakulima Hingilili), highland (WHHO - Water users of Hingilili Highland Organisation) and the sub-catchment (HIBA - Hingilili Irrigation Basin Association). The interviews with the leaders focused on the functioning and responsibilities of these new structures, their interaction with existing arrangements and the challenges they face.

Finally, to triangulate the research findings we organised a two-day feedback workshop with water users and their representatives from Hingilili, as well as Same district department heads and representatives from the PBWO. During the workshop we employed tools to trigger the active participation of water users, including a role playing game on water allocation between farmers from highland and lowland. The role playing game used was the "River Basin Game" developed by Bruce Lankford (see: Lankford et al. 2004).

3.4.2   Study area

Hingilili sub-catchment (about 150km^2) is located in the South Pare Mountains within the Pangani river basin. Figure 3.1 is a detail map of Hingilili sub-catchment showing furrows intakes, villages and climatic zones. The Hingilili river drains into the Mkomazi River, which drains into the Pangani River, which empties in the Indian Ocean. The sub-catchment covers part of eight wards (i.e. Msindo, Maore, Vuje, Bombo, Mtii, Ndungu, Chome and Vudee) in Same district. However, water is only used by the inhabitants of three of the eight wards. In the highland these wards are Vuje and Bombo with four villages (Vuje, Mvaa, Mjema and Bombo). In the lowlands it is Maore ward with four villages (Maore, Mpirani, Kadando and Muheza).

Hingilili sub-catchment can be divided into three agro-ecological zones: a) the forest zone (above 1600m above sea level) comprising a government gazetted forest reserve; b) the highlands (800–1600m above sea level) with mainly subsistence farmers; and c) the lowlands (lower than 800m above sea level) occupied by subsistence farmers and livestock keepers. The area experiences two rainy seasons per year, a long season starting in March and ending in May ("Masika") and the other a shorter season starting around October and ending in December ("Vuli"). The average annual rainfall varies substantially according to altitude: 1000-1500mm/year in the highlands and about 500mm/year in the lowlands.

The Hingilili sub-catchment hosts about 34,500 inhabitants (Tanzania 2002b), many of whom live in the highland (about 18,500 people) the majority being ethnic Pare. Lowland inhabitants belong to a mix of ethnic grouping: Pare, Sambaa and Maasai. Land tenure in the sub-catchment is customary and holdings vary from 0.5 – 5.5ha with an average of 0.8ha (JICA 1984). The main activities are subsistence agriculture based on rainfed and supplemental irrigation, livestock keeping and small agri-businesses.

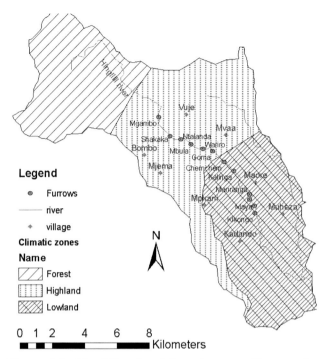

Figure 3.1: Hingilili sub-catchment map, villages using water, their furrow intakes and climatic zones.

Currently, there are twelve irrigation canals, locally called furrows, used by farmers to divert water from the Hingilili river (Figure 3.2). Six furrows (Mgambo, Shakaka, Mbula, Ntalanda, Wariro and Goma) are located in the highlands, with the main crops grown being sugarcane, banana, ginger, yams, cassava and coffee (perennial) and maize, vegetables, beans, groundnuts and sweet potatoes (seasonal). The other six furrows (Rushoto, Mariranga, Kalinga, Maya, Kikongo and Chemchem) are in the lowlands, where farmers grow banana, rice, maize, beans and vegetables. One furrow, Shakaka, from the highland also supplies water to the lowland. Tail water (irrigation water remaining in the irrigation canal at the end of the command area) from Chemchem feeds into Mariranga, while Mgambo farmers now claims that instead of closing the intake in the evening, their water is diverted into Shakaka furrow which carries it to the lowland.

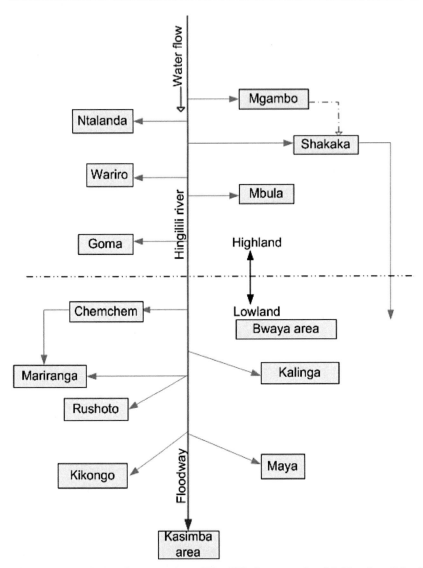

Figure 3.2: Existing furrows along Hingilili river serving highland and lowland areas.

Each of the 12 furrows has a water committee responsible for water allocation, maintenance and conflict management. Table 3.1 presents information on the establishment of the furrows, location, number of users, and command area.

Hingilili sub-catchment experiences water stress during the dry seasons. Increasing water demand arising from natural population growth and changes in cropping patterns (e.g. increase in ginger cultivation in the highlands) make the area a potential hotspot for upstream – downstream water conflicts.

Table 3.1: Establishment and command area of furrows in Hingilili sub-catchment.

Furrow name	Location	Year first established	No of farmers	Command area (ha)
Mgambo	Highland	Around 1900	120	120
Shakaka	Highland	1910	500	1200
Mbula	Highland	1880	100	150
Ntalanda	Highland	1880	216	1300
Wariro	Highland	1880	120	60
Goma	Highland	1880	35	40
Rushoto	Lowland	1910	650	253
Mariranga	Lowland	1910	800	1200
Kalinga	Lowland	1890	900	405
Maya	Lowland	1890	1100	950
Kikongo	Lowland	1920	850	850
Chemchem	Lowland	1920	200	184

3.5 HISTORICAL EVOLUTION OF WATER ALLOCATION ARRANGEMENTS

This section describes how water allocation arrangements in the Hingilili sub-catchment have changed over time and shows that the current water allocation practices are a result of historical developments in and around the sub-catchment. According to historical records, cultivation on the slopes of the South Pare Mountains has been dominated by irrigation systems dating back several centuries (Håkansson, 1998). On the eastern side of the mountains in areas drained by the Hingilili River, many irrigation furrows reportedly crisscrossed the slopes watering fields of maize, beans, sweet potatoes and sugarcane (Håkansson 1998). Since this period (1700 - 1900), agricultural activities have increased in the highlands, as the result of favourable conditions for crop cultivation and increasing market activities. Box 3.1 summarises historical developments in the Hingilili sub-catchment. This section describes the following water allocation practices: within irrigation canals (furrows), between furrows in the highland, between furrows in the lowland, and finally between the highland and the lowland. It will be seen that water allocation in the lowland is most critical, due to more intense water scarcity problems.

Box 3.1: Historical overview of water allocation in Hingilili sub-catchment.

Pre-colonial (before 1880)

South Pare Mountains was divided into small chiefdoms of Mbaga, Mamba, Gonja, Hedaru and Chome (Håkansson 1998). Each of the four sub-catchments on the eastern side of South Pare Mountain was under one chief, from the water source to the lowlands. As population increased in the highlands, more and more people moved to the lowlands for farming and livestock keeping. Trade also developed in the area with the relatively dry western side of the mountain, Usambara and the coast (Håkansson 1998, Sheridan 2002). Some of these markets still operate today, e.g. in Vudee village there is a weekly market attended by farmers from both sides of the mountain. Trade intensified irrigated agriculture and increased the need for allocation arrangements between the furrows. To cope with increased demands the chief initiated rotational water allocation between upstream and downstream farmers.

Colonial period 1880-1967

Formalising of water resource management started in 1914 under German rule (1880-1919) when a first draft of a water rights ordinance was formulated. Cotton estates were established in the lowlands and given first priority over water. The first water rights ordinance was officially proclaimed in 1923 during British rule (1919-1967). By the 1930s, a tax was introduced which was intended to be used by government servants to control the irrigation furrows, prevent wastage by the native farmers upstream and as impartial evidence in cases of dispute. Amendments of the water ordinance occurred subsequently. A department for water development and an irrigation division were introduced in 1959 (Burra and Van den Heuvel 1987).

Post-independence of Tanzania: 1967- 1990

A complete overhaul of the economic system through African socialism and self-reliance in locally administered villages (Ujamaa villages) through a villagization program was implemented in 1973-76. The chiefdom system was formally abolished and locally constructed irrigation furrows became village government property. In 1974 the government put in place a new Water Utilization (control and regulation) Act No. 42 that spelled out procedures for granting statutory water rights with priority given to domestic, livestock, irrigation, industries, hydropower, transport and recreation. The 1974 Act and its subsequent amendments of 1981 set the foundation for water management along hydrological boundaries and mainland Tanzania was divided into nine river basins (Sokile et al. 2003, Komakech et al. 2011b). The government also continued with the concept of "Modern Village Irrigation Scheme" originally designed by the British to decongest the highlands (including the Pare Mountains).

NGO and related development: 1990 – 2004

In 1991 the Pangani Basin Water Board (PBWB) is established. PBWB carried out an inventory of water users in the basin and established a protocol for issuing water rights and setting tariffs. It started registering water users and issuing provisional water rights. In 1991 a Water Policy was introduced which focused mainly on providing clean and safe water. In 2002 the government revised the water policy. The new policy objective was to develop a comprehensive framework for promoting the optimal sustainable and equitable development and use of water resources for the benefit of all Tanzanians. In the Hingilili sub-catchment conflicts over water sharing emerged and NGOs became involved. In 2002, the Traditional Irrigation Improvement Programme (TIP) established a water user organisation. In 2004, IUCN and PAMOJA Trust (a local NGO) assisted highland farmers to create a water user organisation to manage all six furrows in the highland. IUCN and PAMOJA also established an apex organisation.

Formalised institutional structure (after 2004)

The process started of setting up sub-catchment fora for water allocation in sub-catchments of the Pangani basin, Hingilili being part of the Mkomazi catchment forum. The new Water Act of 2009 allows for granting water use rights, with prioritisation of water for basic human needs and the environment, and subject to social and economic criteria.

3.5.1 Water allocation and management within irrigation furrows in the Hingilili sub-catchment

All furrows in the Hingilili sub-catchment follow the same procedure for allocating water to their members. In the pre-colonial era water distribution within a furrow was decided by a member of the family or families who first constructed the canal, and who were also considered clan leaders. Nowadays, every furrow branch (secondary canal) has a branch committee which is responsible for preparing the allocation schedule between farmers of that particular branch. Furrows in the highland have between two and four branches while those in the lowland have more (eight to 18). To get a water turn a farmer (a) must have participated in maintenance, (b) must have paid the seasonal fees, and (c) must attend furrow meetings. The furrow committee is responsible for the water allocation between branches of that furrow, particularly during the dry season or during other periods with low flows. The branches are allocated a time slot within the rotation schedule. Each branch has a water distributor (Mgawamaji) who enforces the agreed allocation schedule - who gets water at what time and for how long.

The way water is allocated within a furrow has not changed significantly but ownership and responsibilities have changed over time. All furrows were originally owned and managed by the families and clan members that established it. After independence, the furrows became village property with the village government acting as custodian. Any member of the village could become member of the furrow system. The water distributor is currently elected by the village water committee (cf. Kemerink et al. 2009). For many of the furrows in the Hingilili sub-catchment, the elected water distributors are still descendants of the founding families.

Although the furrow committee defines in detail a schedule for water allocation, there are no division gates to fully control water flows to fields so it becomes difficult to strictly enforce the allocation schedule and to prevent water theft. In the pre-colonial days chiefs were the highest body to resolve water conflicts; during the German colonisation the Akidas (African or Arab administrator of a section of a district) played this role; the British reinstated the chiefs in 1919. After independence, the government abolished the chiefs and their leadership role was taken up by the village government (Mwakalukwa 2009). Currently the water distributor is the first person responsible for dealing with water conflicts. When the distributor cannot resolve the issue, it is taken to the furrow committee. For conflicts that cannot be settled by this committee, the village government refers the parties to the ward office.

3.5.2 Water allocation between furrows in the highland

Currently there is sufficient water to supply all furrows in the highlands during the wet season. In the dry season water allocation between furrows is done on a rotational basis, starting from Tuesday to Saturday but during periods of extreme scarcity it starts on Monday. On Sunday, water is reserved for the environment (cf. Mul et al., 2011). Each furrow, starting with the most upstream, takes turns to divert water (Mgambo, Ntalanda, Wariro, Shakaka, Mbula and then Goma respectively). However, the recent rehabilitation of the furrows has triggered new dynamics into the system, as now more water can and is being diverted by the furrows in the highland zone. In 1992, TIP (Traditional Irrigation Improvement Programme) in collaboration with the local government improved two furrows (Ntalanda and Mbula furrow). In 2009, furrow committees in the highland received funding from the Participatory Agricultural Development and Empowerment Project (PADEP) and used it to rehabilitate all six highland furrows (lockable intakes were installed and some sections of the canals were lined).

3.5.3   Water allocation, conflict and management between furrows in the lowland

In the lowlands, during periods of low flows water is allocated on a rotational basis between the furrows. This local system has been threatened by developments in the sub-catchment, such as increased demand due to population growth; degraded sub-catchments due to the villagization policies (i.e. compulsory resettlement of people by the government into designated villages or collective settlements to improve service delivery and to enhance agricultural production) (cf. Kikula 1997: 140); improvements of the furrow intakes in both the highland and lowland; and the increased cultivation of high water consuming crops such as ginger and sugarcane in the highlands. The expansion of agricultural lands into protected areas, requiring additional water, has aggravated water scarcity, and has increased the occurrence of water conflicts.

According to farmers, water conflicts first emerged in the 1970s following years of droughts, when Mariranga farmers would widen their intake during the dry season. There were more problems when the 1978 floods changed the river course into Mariranga furrow leaving Kikongo, Rushoto and Maya without water. In 1992 the local government in collaboration with TIP reconstructed the damaged intake of Mariranga, Rushoto, Maya and Kikongo furrows. However, the new intakes made permanent the inequity of water sharing between the furrows (e.g. after rehabilitation Mariranga furrow would extract more water); this is why the intakes were later destroyed by farmers. Farmers claim that the intakes were poorly designed such that no water could flow through some gates (e.g. the Maya intake could not divert low flows), while other furrows would divert most of the river flow (e.g. Mariranga furrow). Other intakes were prone to silting (e.g. Rushoto furrow), causing local flooding at the site. In addition to constructing the intakes, a 4.5 km floodway was constructed by TIP to control flooding of the sub-village of Kadando during the rainy season. After constructing the floodway, the plain land (Kasimba) originally used for livestock grazing became frequently flooded. This fertile land was subsequently occupied by farmers who started growing rice (about 200ha belonging to 250 farmers).

Over time conflicts over water distribution between the furrows increased and the Maore Ward Executive Officer (under whose charge most furrows fall) was unable to resolve most of them (Mwakalukwa 2009). An example of this was the fierce water conflict that erupted in December 2000 between two furrows in the lowland (Rushoto and Maya furrows). The executive officer and the water committee linked to the village government both failed to adequately resolve the conflict. The conflict reached a critical point whereby government police and Same District authority had to restore order. After two years of negotiation, involving the District Commissioner, Same District Council, PAMOJA, TIP and lowland farmers, it was agreed to form an umbrella organisation called MUWAHI (Muungano wa Wakulima Hingilili) to manage water allocation between the furrows in the lowland. MUWAHI was registered with the Ministry of Home Affairs in 2002 (Box 3.2).

Box 3.2: Muungano wa Wakulima Hingilili (MUWAHI).

Members of the MUWAHI committee comprise of elected representatives of the six lowland furrows. Each furrow conducts an election starting at the level of a branch canal. In total five representatives are elected from each branch canal to form the furrow assembly. The furrow assembly then elects five representatives to represent them on the MUWAHI committee. The chairman, vice-chairman and secretary of MUWAHI are elected in a general assembly attended by lowland farmers. The current chairman of MUWAHI is the chairman of the Mariranga furrow, while the secretary of MUWAHI is the vice chairman of the Mariranga furrow. The combination of a good-functioning intake and a strong representation at MUHAWI give Mariranga furrow farmers an edge over farmers from the other furrows, who feel that their access to water is being constrained by the Mariranga intake structure.

Each of the six furrows also formed a water user group which were registered under the Rural Cooperative Society Act. The floodway water users of Kasimba are not part of MUWAHI - because the land belongs to several villages, their water issues are reportedly discussed at ward level. Farmers in Kasimba do not get water during periods of low flows. Because of their reliance on "matupio"Kasimba farmers are the losers in times of water scarcity and the winners when there is excess flows. In December 2009, the Rushoto Maya and Kikongo farmers got funding from PADEP to relocate the Rushoto Maya and Kikongo furrow intakes back to their original locations and also lined some of the canals.

Water allocation in the dry season between furrows is now supervised by MUWAHI through its sub-committee for water, agriculture and the environment. The sub-committee is responsible for collecting the dry season cropping calendar (May – October) from each of the six lowland furrows (we found two such dry season calendars, namely for the years 2007 and 2008). During the period May – October four furrows (Rushoto, Mariranga, Maya, and Kikongo) are only allowed to allocate water to three branch canals; the remaining branches are not to be irrigated. Chemchem and Kalinga furrows were not rehabilitated by TIP - the two furrows are believed to loose nearly 50% of the diverted river water before reaching the fields. To compensate the farmers of these furrows for this disadvantage, they are allowed to divert water throughout the week even during dry seasons. The other furrows often enter into rotational allocation agreements that vary with the level of water scarcity. In case of extreme scarcity the remaining furrows are paired: Rushoto and Mariranga furrows irrigate for a day and the next day water is for Maya and Kikongo furrows.

MUWAHI often advises furrows to cultivate only part of their command area (e.g. Mariranga furrow with 18 branches are advised to cultivate the area served by eight branches only). In reality, however, the agreed cropping calendar is not completely adhered to and farmers tend to cultivate their entire land area, whatever has been agreed. In addition, the intake of Mariranga furrow is the only one able to abstract low flows, while the other intakes are prone to silting. Farmers in the Mariranga furrow can therefore grow rice year around, while others cannot, which exacerbates existing inequities in water access.

3.5.4   Water allocation and management at the sub-catchment level

During the pre-colonial era, trade intensified agriculture and increased the need for water allocation between furrows using the Hingilili River. As a result, the chief initiated a rotational system between the highland and lowland farmers. The present day water allocation system between highland and lowland is still based on this age-old rotation system. During the day farmers in the highland use water for irrigation (from 6 AM to 4 PM), after that they are supposed to close their furrow intakes and leave the water for lowland farmers. Furrow intakes, however, should not be completely closed, as some minimum flow must be maintained for domestic and livestock needs. On Sundays water is left in the river to sustain the environment and wildlife (cf. Mul et al. 2011).

During German rule cotton estates established in the lowland were issued water rights. This system severely affected smallholder farmers in the lowland, while the highland farmers could continue with their practices, and were encouraged by the Germans to grow coffee, creating a source of income that was taxed by the colonial rulers (Burra and Van den Heuvel 1987).

The villagization programme following independence of Tanzania further alienated the highland from the lowlands putting them in different administrative units. In the 1970s, the highland became part of Gonja division and the lowland became part of Ndungu division. Water conflict resolution at the sub-catchment level initially was very weak; highland villages fully exploited the water resources and left the lowland villages with limited irrigation water. In addition, land areas previously unused were subjected to deforestation for firewood, timber and clearing for new farmland. To solve water conflicts between the highlands and lowlands, and to ensure good relations between the two areas, a neighbourhood committee (known as Kamati za Ujirani Mwema) was established in 1994, which consists of two parts - the highland and the lowland (PAMOJA 2004). The committee comprises of furrow leaders, village government representatives, the divisional secretary and the ward councillor. The meetings are chaired by the Division secretary of Ndungu and the Division secretary of Gonja acts as the water committee secretary. The good neighbourhood meetings still take place to this day and issues discussed include matters of defence and security, as well as water. Although the neighbourhood committee does not play a pro-active role in day-to-day water management, it comes into action when a water conflict arises between the two divisions (Gonja and Ndungu).

Currently, the lowland farmers pay a representative to go upstream to close the furrow intakes according to the rotational schedule. This task has recently become more difficult because the intakes are now locked so the representative must first look for the intake keys in the villages. Another challenge to the rotation system is the increased cultivation of ginger and sugarcane which has made highland farmers reluctant to close their furrows at 4PM. Ginger was introduced in 2006 by a local NGO (Faida Mali), and has since then been cultivated by many farmers because of its high market value. Ginger requires supplementary irrigation for at least nine months of the year. Its increased cultivation in the highland means increased water use resulting in more pronounced water scarcity in the lowland, and a potential for renewed conflict in the sub-catchment.

3.6 GOVERNMENT AND NONGOVERNMENTAL INTERVENTIONS SINCE 2003

The previous section chronicled institutional arrangements up to about 2003. This section briefly describes recent attempts by PBWO and local and international NGOs to reconcile the emerging problems between the highland and lowland areas.

3.6.1 Formation of a sub-catchment apex organisation

Through a project called "Dialogue on Water", IUCN, PAMOJA, PBWO and Same District council were from 2003 to 2005 involved in the Hingilili sub-catchment (PAMOJA 2004, Tack 2006). The project used Same District council's experience with conflict resolution in the Hingilili lowland area to raise awareness among highland farmers, to create an organisation similar to MUWAHI in the highland called Water users of Hingilili Highland Organisation (WHHO), and also to create an apex organisation called Hingilili Irrigation Basin Association (HIBA, Box 3.3) that would link the lowland (MUWAHI) and the highland (WHHO). The creation of HIBA was in line with the National Water Policy of 2002 but may also be considered an attempt to nest water institutional arrangements (Ostrom and Gardner 1993). WHHO would be responsible for the water allocation between the furrows in the highlands, while HIBA would oversee the implementation of the agreements between the highland and the lowland. HIBA would also be responsible for enforcing rules on source protection as well as identify market opportunities for the agricultural produce. HIBA is not officially registered with PBWO and ever since the "dialogue on water" project ended HIBA has not functioned and has effectively ceased to exist.

Box 3.3: Hingilili Irrigation Basin Association (HIBA).

The HIBA committee is made up of representatives from MUWAHI and WHHO- each organisation being represented by 12 members. These representatives elect the chairman, vice chairman, secretary and treasurer. At the time of the research, the chairman and secretary of HIBA were the chairperson and the sectary of MUWAHI respectively. The vice chairman and treasurer were from WHHO. HIBA was supposed to be financed from the user fees collected by MUWAHI and WHHO. However, MUWAHI and WHHO have not contributed any funds as they claim none of the furrows members are willing to financially support HIBA.

Although MUWAHI is actively following up the agreed water allocation between the highland and the lowland, HIBA and WHHO are not directly involved. Recently, the HIBA chairman who is also MUWAHI chairman, tried to institutionalise an arrangement whereby lowland farmers would pay WHHO to engage the highland farmers in closing their furrow intakes at 4PM but the idea was rejected by MUWAHI officials. The latter argue that it is the responsibility of the highland furrows to close their intakes according to the old agreements. The highland farmers, however, object and argue that some intakes are far from the villages and they need incentives to close them. A similar arrangement is practiced in the neighbouring sub-catchment of Yongoma serving Ndungu irrigation scheme, where lowland farmers currently pay highland farmers to close their gates at 4PM and it is reportedly working well. In addition the lowland farmers currently pay some fellow irrigators to close the upland intakes.

3.6.2   Linkages between state-led water rights reforms and local practices

According to the Water Utilization (Control and Regulation) Act of 2009 all water users, individuals or groups, are obliged to apply for and obtain a water use permit (originally called a water right). Water use permits are issued and administered by the Basin Water Board. The permit indicates the purpose and amount (l/s) granted to the holder. Once granted the permit holders are required to pay an annual water user fee that is determined by the type of use and the extraction rate stated on the permit. In sub-catchment areas where the volume of water is inadequate to satisfy all permits granted, the Basin Water Board may review and revise the use, diversion, control and allocation of water in the area. Any non-domestic users, who had beneficially used water from a source for an uninterrupted period of more than five years without a permit were entitled to a water use permit provided they applied within two years after the commencement of the Act (Tanzania 2009). Customary rights held by a community in a watercourse are recognised and considered of equal status to a government granted water right. Effectively these operate as granted right and are to be recorded by a basin water board in favour of the user group. The granted right can still be governed by customary law in respect of any dealings between persons using the water source and may be subject to an annual use fee. Customary right holders and other users organisations, such as associations and cooperative societies, may all apply to the Basin Water Board for water use permits.

Based on the state law, Pangani Basin Water Board is in the process of granting water rights to the furrow groups in the sub-catchment. The twelve irrigation committees all have paid the mandatory application fee (Tsh. 40,000 per furrow irrigation committee, about 30 USD) but at the time of the research they had not completed the application procedure. This notwithstanding, the furrow committees nevertheless have started paying the annual water use fee (a flat rate of Tsh. 35,000 for each small scale irrigator, about 26 USD) to the basin water board since 2008. This charge forms part of the annual membership and entrance fees contributed by members and new irrigators for each furrow, respectively. By supporting the creation of user associations in the lowland and highland (MUWAHI and WHHO) and an apex organisation (HIBA) in Hingilili sub-catchment, the Pangani Basin Water Board believes that the new arrangements will provide an effective interface between local arrangements and the basin water board. In addition, the Board expects that these structures will simplify the collection of the annual water user fees. However, the overlap between local and state-led institutional arrangements as shown in Figure 3.3 is complex and messy (cf. Mehta et al. 1999). It is therefore unlikely that the state created structures will simplify the collection of the annual water user fees.

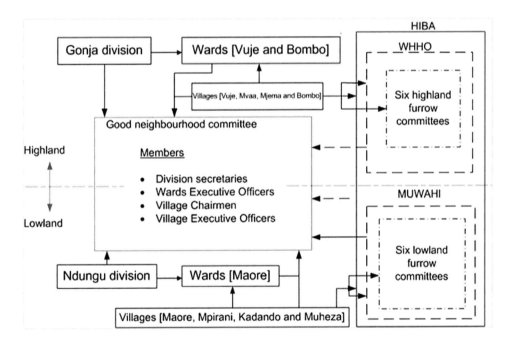

Figure 3.3: Institutional overlap at various levels: Hingilili furrow committees, highland and lowland organisation (WHHO and MUWAHI), sub-catchment organisation (good neighbourhood committees and HIBA) and administrative offices (Divisions, wards, and villages). Dotted arrow means weak interaction.

3.7 DISCUSSION: INTERFACE AND IMPACTS OF FORMALIZATION

The preceding sections described the evolution of water management in Hingilili sub-catchment. In this section we discuss the impacts of recent attempts by the Pangani Basin Water Board, Same district local government and NGOs (both local and international) to regulate water management in Hingilili sub-catchment. Water user institutions that link upstream and downstream users were created and furrows rehabilitated. These state-led developments were not only a response to local water conflicts in the sub-catchment but may also be viewed as a concerted attempt to implement the national water policy in the Pangani basin. The stated objective was to achieve equitable water allocation and sustainable management of the water resources. However, the historical review indicates that the new water law includes elements first introduced by the colonial administrators as early as 1923. The British introduced the registration of water rights, mainly to limit water use among the natives and to secure access for European settlers (Komakech et al. 2011b). The colonial water law has been amended in subsequent water acts and more recently in 2009, but its regulatory aim continues to focus on the requirement of all water users to register and to obtain a permit. In addition to registration, an annual water user fee was introduced in the National Water Policy of 2002 and passed into law in the Water Act of 2009.

We find that the state-led reforms have not changed the day-to-day local water allocation rules significantly. Current water allocation practices in Hingilili sub-catchment are still based on procedures established in the pre-colonial era. Within a furrow access to water is based on farmers' participation in maintenance and attendance at meetings. The age-old "day-night" turn taking between the highland and lowland is still in place. The allocation rules are well understood by everyone and there is a shared expectation of all users to cooperate. However, recent infrastructure rehabilitation supported by NGOs has negatively affected the arrangements between the highland and lowland farmers. For example, following the rehabilitation of furrows in 2009 and the introduction of a new high value crop - ginger - in the highland, more water is now being used upstream. In addition, the highland farmers now claim that they need to be paid to close their furrow intakes at 4PM, while lowland farmers maintain that it is the responsibility of the former to leave sufficient water for downstream use. People with location advantage tend to ignore or reject established rules and norms.

The above observation does not mean that local water management structures have not evolved over time. Furrow ownership and responsibilities have changed. Furrows used to be owned by the founding clans but are now village government property. In addition, instead of the founding families it is now periodically elected furrow committees and water distributors that are allocating water to individual farmers. Chiefs used to be the highest body to resolve water conflicts but now this is the responsibility of the ward office. However, the graduated process of water conflict resolution has not changed: water conflicts that the water distributors failed to solve are still referred to the next highest level (furrow committee) as it was done during the era of the chief.

The persistence of local arrangements in Hingilili sub-catchment fits well with the concept of institutional bricolage. The fact that the existing institutions still draw from pre-colonial practices is illustrative of the point that institutions are constructed from existing logics, rules and norms (Cleaver 2002, Galvan 2007). The neighbourhood committee (Kamati za Ujirani Mwema) emerged to fill the void left when chiefs were abolished. This highlights how socially embedded institutions are often adapted to solve new challenges. Presently the neighbourhood committee operates on an ad hoc basis and mediates water conflict between the highlands and lowlands, a role previously played by the chief.

Hingilili farmers have largely been able to cope with changing conditions, such as population growth and changing state policies by creatively reusing local institutions. Nevertheless, frequent floods particularly in the lowland around 1978-9 and subsequent infrastructure rehabilitation by TIP in the early 1990s did cause water conflict. The reconstructed furrow intakes created an unfair advantage for some furrows and thus constrained others. This unequal distribution fuelled water conflicts among the lowland furrows. Setting up a water user organisation (MUWAHI) has to some degree reduced the level of water conflict among lowland furrows; but it also created winners, such as Mariranga furrow whose interests are much better represented at all levels compared to other furrows. Fierce competition has been pushed further downstream in the river basin. The construction of a floodway led to the emergence of floodway farmers and livestock keepers in the Kasimba area. Kasimba farmers' claim to be recognised as legitimate water users is contested on the grounds that their land is not irrigated by a furrow and that the water they rely on is excess water not needed upstream.

Formalisation in the Pangani basin included the registration of water users, the issuing of volumetric water use rights to individual irrigators and groups, and the creation of water users organisations at various levels. The two major aims of Tanzania's water reforms have been to improve economic efficiency and restore order and rationality over water use. Water use rights are expressed in fixed volumetric terms, regardless of wet and dry season and wet and dry years. The permit, however, does state that there is no guarantee that the quality and quantity of water referred to will be available. The Pangani Basin Water Board started with registering water users at the level of furrow committees, but the Water User Associations (MUWAHI, WHHO and HIBA) have not yet been registered. None of the 12 furrows had been issued a (provisional) water right at the time of research.

The concept of nesting institutional arrangements (see: Ostrom 1993, Ostrom and Gardner 1993) at first appeared relevant but raises several questions. At what level should the state create formal arrangements? How should new institutions be linked with the existing local arrangements? How have the pre-existing institutional arrangements resolved the issue of nesting and of scale? The Pangani Basin Water Board together with some NGOs invested significant amounts of resources (time and money) in creating nested governance structures in Hingilili sub-catchment: HIBA was created to link MUWAHI and WHHO, while ignoring the existing neighbourhood committee. Several training sessions and exchange visits for the leaders of MUWAHI, WHHO and HIBA were organised. However, the structure that was supposed to serve as the apex organisation that would manage water allocation between lowland and highland (i.e. HIBA), is currently ignored by lower-level structures. This is exemplified by MUWAHI engaging directly with individual furrows in the highland. In addition, the neighbourhood committee that comes to life whenever there is conflict between highland and lowland, has taken on the conflict resolution role that the Pangani Basin Water Board assigned to HIBA. HIBA is currently a redundant organisation and unnecessary, despite the fact that upstream-downstream contradictions have not yet been fully resolved. More importantly, the fact that the organisational structure is nested seems to be designed to provides opportunities for elected representatives to play multiple roles at multiple levels – and this is exactly what happens in practice. The chairman and secretary of Mariranga furrow, for example, also act as chairman and secretary of MUWAHI and HIBA respectively. The higher level organisations have been hijacked by a handful of individuals that aspire to secure and extend their power base, for example to consolidate control over access to water for the furrows in which they have a vested interest. The failure of the apex organisation to function as designed shows that many other irrigators do not agree with this type of concentration of institutional opportunities and resources that seems to heighten existing inequities and power differences. The inability of apex organisation to function as designed also highlights ambiguities of institutional overlaps and linkages between local and state forms at a sub-catchment level, described by Metha et al (1999) as the 'messy middle'. Here the institutional arrangements may be highly contested and beset with ambiguity or even reinterpreted (Mehta et al. 1999).

Water institutions operating at the scale of Hingilili sub-catchment appear to be ideally positioned to manage and allocate water resources in the 'messy middle'. To grant these types of organisations the right to issue water use permits and to levy water user fees could potentially reduce transaction costs. This is so because they would build on existing institutional arrangements, making it more acceptable for irrigators to negotiate day-to-day water allocation in the same way they have been allocating water in the past without too much outside interference. However, in the case of Hingilili, HIBA is not operational; the only functional water institutions beyond the level of the individual furrow appears to be MUWAHI and the neighbourhood committees. The institutional arrangements of WHHO and HIBA were not sufficiently embedded and their linkages with existing arrangements were weak. To be able to support and create an effective sub-catchment organization, the Pangani Basin Water Board should build upon existing structures such as the neighbourhood committees rather than introduce new arrangements. However, this is by no means a guarantee for success - institutional arrangements are messy and often get reinterpreted and re-negotiated at the local level.

3.8   CONCLUSIONS

In this chapter we explored the effects of state-led formalisation of water allocation on local water management practices in Hingilili sub-catchment, Tanzania. Uncoordinated infrastructure improvement and change in cropping patterns are a source of instability in the sub-catchment. However, we find the local age-old rotational allocation that was developed based on the principle of good neighbourliness between highland and lowland farmers still partly in use. Access to water within a furrow is based on local rules and norms that have evolved over time. Most of these rules stem from the pre-colonial era. When chiefs were abolished and furrows became village property, the local arrangements were reinterpreted and evolved into a neighbourhood committee with a much wider mandate than only water. This committee has successfully mediated conflicts between highland and lowland residents, including those related to water. Clearly, actors draw from existing logics and rules to craft new institutions (Cleaver 2002, Galvan 2007).

In sum, the state-led formalisation of water allocation and management has so far had little impact on actual day-to-day water allocation practices in Hingilili sub-catchment. The Pangani Basin Water Office is still in the process of registering and issuing water rights. Currently farmers seem interested in acquiring a state-sanctioned water use right because they believe it will add legitimacy to their existing claims to irrigation water. To date no furrow groups in Hingilili have been granted formal water rights but, remarkably, all furrow groups do pay the annual water user fee.

We find that the concept of institutional nesting failed to work in the sub-catchment. Whereas the creation of the lowland water users association succeeded in minimising water conflicts between farmers there, the newly created highland association and sub-catchment apex organisation both failed. The ready explanation is that the latter organisations were not properly linked to existing institutional arrangements. This is likely to have been correlated with the fact that there is no need nor incentive for upstream furrows in the highland to engage with their lowland counterparts, because of their location advantage - there was therefore a paucity of institutional resources to build on. Hence the relevance of appreciating the institutions that do exist at catchment level – including and in particular the neighbourhood committees. Local water management arrangements could have been strengthened by the government if it would have recognised the neighbourhood committees and mandated it to also mediate water conflicts between highland and lowland.

The concept of bricolage sensitises the need for new institutions to be sufficiently embedded into existing local practices to succeed. Even then, this does not guarantee it will lead to equitable access and sustainable water management. Any new institution will be subjected to processes of construction and de-construction by the actors involved – i.e. bricolage. It is possible and very likely that powerful actors will reject established rules and jump on opportunities created by outside interventions to gain control over water. The hydraulic position of the various actors (upstream or downstream) adds a complicating dimension to these institutional dynamics that makes things more difficult, and may be considered as a motor or driver for institutional innovation. It may be hypothesised that institutional innovation in a catchment gradually crawls upstream as the needs for coordination grows accordingly.

Chapter 4

POLYCENTRISM AND PITFALLS - THE FORMATION

OF WATER USERS' FORUMS IN KIKULETWA

CATCHMENT, TANZANIA[6]

4.1 ABSTRACT

Catchment forums have been proposed as appropriate arenas that allow actors to
dialogue and participate in decision-making related to water management. However,
conditions found in many developing countries present significant implementation
challenges. River catchments typically cover large numbers of administrative districts,
host diverse stakeholders and have institutional arrangements with overlapping
jurisdictions of state-led and locally created institutions. Institutional nesting has
been proposed as a companion to the catchment forum concept as this may help to
integrate local arrangements. However, the creation of a polycentric or nested
governance system is not straightforward and raises questions of how to coordinate
diverse semi-autonomous lower-level units. This chapter highlights the difficulties of
designing catchment forums in an African context. The chapter describes and
analyses an attempt by the Pangani Basin Water Board to create a catchment forum
in the Kikuletwa catchment, Tanzania. The process of developing this forum faced
many problems. Zoning the catchment into sub-catchments produced water users

---

[6] Based on Komakech and Van der Zaag, 2013. *Water International*, 38(3), 231-249.

associations that were weakly embedded in local structures. Resolving the problem of administrative boundaries and institutional fit while integrating customary arrangements with the state-led governance structure requires careful analysis of local structures.

4.2 INTRODUCTION

A basin/catchment is considered to be closing when commitments for domestic, industrial, agriculture or environmental uses cannot be met for part of a year and closed when these commitments cannot be met over the entire year (Molle et al. 2010). This situation often intensifies competition and sometimes leads to violent conflict over water (Komakech et al. 2012e). For such catchments institutional arrangements that can coordinate the use of water across scales and levels are needed. Scale here is defined broadly as the spatial, temporal, quantitative or analytical dimensions used to measure and study objects and processes (Gibson et al. 2000). The concept of integrated water resources management places the participation of water users in decision making high on the agenda as it is thought to lead to better decision making and coordination.

Catchment forums have been proposed to (Warner 2005, Faysse 2006, Robinson and Smith 2010) provide spaces that allow water users to engage in meaningful dialogue and participate in decision making. They are multi-stakeholder platforms that involve representatives of different use sectors (agriculture, domestic use, hydropower etc.), as well as upstream and downstream actor groups. As platforms, they structure an arena where actors with competing interests meet and seek consensus on issues such as water allocation, negotiation of new rules, and resolution of conflicts (Warner 2005, Swatuk 2008).

Although catchment forums and stakeholder platforms are now important catch phrases for many international donors and governments, their implementation has fallen short of expectations (Manzungu 2002, Waalewijn et al. 2005, Faysse 2006, Wester et al. 2008). South Africa and Zimbabwe have experimented with catchment management agencies and catchment councils respectively but so far these institutions have failed to achieve stated policy goals (Dube and Swatuk 2002, Manzungu 2002).

The challenge is how to better organise catchment forums in practice. Scholars have proposed that large catchments be decomposed into smaller, distributed and autonomous decision making sub-units that constitute simultaneously a whole and a part (Andersson and Ostrom 2008, Lankford and Hepworth 2010, Ostrom 2010). The assumption here is that collective action problems faced by large groups are decomposable into smaller problems solvable by small semi-autonomous groups (Marshall 2008). Water allocation and conflict can then be resolved within the subunits and between them (Lankford and Hepworth 2010). Modularising catchments in this way simplifies monitoring as only a few points need to be checked. In addition, the sub-units can be based on local pre-existing institutional arrangements. As multiple and overlapping decision making centres retain considerable degrees of autonomy, this creates a nested polycentric governance structure (Ostrom et al. 1961, Ostrom 1990, Marshall 2008, Ostrom 2010). The feasibility of nesting is inspired by the positive evidence available on local capacity to self-organize and craft effective institutions for solving collective action problems (Wade 1988, Ostrom 1990, Ostrom 1993, Komakech and Van der Zaag 2011). Even if this is possible, however, a number of serious concerns arise. There is no guarantee that a polycentric system will be able to find optimal combinations of rules at the various levels they operate at (Ostrom 1999); whereas local institutional arrangements may be enduring, they are not necessarily equitable; and polycentric systems may in fact provide an opportunity for powerful actors to strengthen their networks and sustain or even increase inequity in water access.

Catchment forums have been a feature of recent water management reforms in Tanzania. Linked to a strong policy of decentralisation by devolution, Tanzania formulated a Water Policy (2002) and enacted a new Water Act (2009) that provide for active participation of water users (Tanzania 2002a, Tanzania 2009). Nine basin boards have been created which are overseeing the establishment of lower level structures including catchment forums and water users associations. In the Pangani basin, the basin water board and development partners (both local and international NGOs) are piloting catchment and sub-catchment forums. The establishment of these lower structures aims to address emerging water conflicts.

In this paper we explore the process and formation of a catchment forum in the Kikuletwa catchment of the Pangani basin. We observe the challenges faced in designing catchment forums following a nesting approach in a river catchment with a diversity of actors and institutional arrangements.

Section 4.3 provides a review of the catchment forum concept and the concept of polycentric governance. Section 4.4 introduces the case study catchment and describes the institutional environment focusing on state-led and locally evolved arrangements. Section 4.5 presents the process and formation of sub-catchment water users associations in the Kikuletwa catchment. Section 4.6 highlights some of the challenges and pitfalls and section 4.7 draws conclusions on the feasibility of a catchment forum as well as that of polycentric governance.

4.3   CONCEPTUAL REVIEW OF CATCHMENT FORUMS

The concept of a catchment forum draws from collaborative and communicative rationality theory, and concerns a process whereby two or more actors pool their appreciation and capacities to address a problem that they cannot solve individually (Waalewijn et al. 2005). Three characteristics of a forum can be identified, namely voluntary participation of the actors, direct face to face interactions among the representatives, and mutual consensus and agreement on action strategies by all the parties affected (Brody 2008). The concept is based on the assumption that as actors start talking, a process of learning takes place in which power gaps and institutional hindrances are broken down; as a result, actors' perceptions and definitions of the problem change and converge (cf. Poncelet 2001). Actors may revise their preferences in light of new information made available to them (Neef 2009). Thus the belief is that once a catchment forum is established, equitable allocation and management of the water resources can be realised, as it provides an arena where users have equal opportunity to debate, rationally consider and reach consensus on water management problems at stake (cf. Brody 2008). This would make it suitable for water stressed catchments.

However, despite the idea of stakeholder participation having been in the water management domain for some time, empirical cases of meaningful participation, especially by poor water users, remain rare. Many scholars report that actors' participation in decision making and management often remains limited to consultation (Cleaver 1999, Wester et al. 2003, Neef 2008). Warner (2005) argues that although actors do acquire new information and ways of thinking from participating, collaboration does not necessarily follow. Collaboration implies situations where decisions are jointly made, power is shared, actors undertake collective action and accept the outcomes of their decisions (Brody 2008). Catchment forums face an additional challenge, namely that they often comprise relatively large areas drained by several tributaries falling in different administrative areas. Some tributaries may experience higher seasonal variability than others, and may not connect with the main stream during some months in the dry season. In such cases it is more difficult for users in different tributaries to acknowledge their hydraulic interdependencies.

Hence nesting new catchment forums with lower level self-organised arrangements has been proposed to overcome some of the coordination problems faced by many groups of users dispersed over a large area (Andersson and Ostrom 2008, Lankford and Hepworth 2010, Ostrom 2010). This, in theory at least, should allow the smaller, self-organised organisations to become part of a larger system without losing much of their identity and autonomy.

However, the success of this approach depends on the ability to identify suitable subunits as well as the mechanisms and services needed to support water dialogue within and between the subunits (Neef 2009). This is by no means simple in catchments with diverse number of actors, who have also developed different systems of water allocation and management.

Despite the perceived potential benefits of polycentrism, empirical evidence is lacking to prove its success. The biggest challenge is the effective coordination of fragmented organizations that lack a central focal point (Sovacool 2011). As each subunit may make its own distinctive rules, it is likely that a variety of governance arrangements will emerge to interact horizontally (i.e. across the same level) and/or vertically (i.e. across different levels of organisation) (Marshall 2008). It is unlikely that all these arrangements will be consistent with government policy objectives.

4.4   CASE STUDY: KIKULETWA CATCHMENT

In this section we introduce Kikuletwa catchment and then describe the attempts by the Pangani Basin Water Board and its collaborating partners to institutionalize a catchment forum.

4.4.1   Research methods

This paper is based on research conducted between August 2008 and September 2010 in the Kikuletwa catchment, Tanzania. Data on local institutional arrangements are derived from related research conducted on the emergence of river committees in the Themi sub-catchment (Komakech and Van der Zaag 2011), cooperation between estates and small-scale irrigators in the Nduruma sub-catchment (Komakech et al. 2012a) and on water right enforcement in the Weruweru sub-catchment. Information on the catchment forum process was collected through interviews and discussions with key actors (farmers, Pangani Basin Water Office (PBWO), Pamoja, and SNV staff), field visits, mapping, and observations of the catchment forum process. The first author participated in seven workshops organised by PBWO on the catchment forum. The paper also draws from grey literature obtained from PBWO, and Pamoja Trust and SNV Arusha, a local and international NGO respectively.

4.4.2   Biophysical and socio-economic context

Kikuletwa catchment covers the northwestern part of the Pangani River Basin (Figure 4.1). The catchment area measures approximately 6,650 km^2. It covers parts of six administrative districts and comprises 80 administrative wards. It is drained by fifteen major rivers originating from Mount Meru and Mount Kilimanjaro. These rivers join to form the main Kikuletwa river before entering Nyumba ya Mungu reservoir downstream.

The water users include small scale subsistence farmers, two cities (Arusha and Moshi), a number of small towns, large scale export/commercial farms, pastoralists, mines and tourist facilities. Kikuletwa River is the main source of water of the Nyumba ya Mungu reservoir that regulates water for electricity production further downstream.

With the increase in population, people living along the slopes of Mount Meru and Mount Kilimanjaro now intensively farm their land. Farmers utilize most of the waters from streams/rivers originating in the highlands. As a result the volume of water flowing from Themi, Nduruma, Malala, Usa, Sanya and Kware rivers has decreased drastically. Some sections of the main Kikuletwa River now periodically dry out.

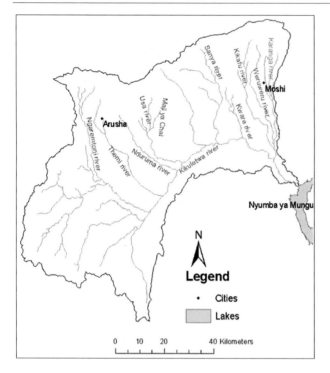

Figure 4.1: Kikuletwa river catchment, its major tributaries and Nyumba ya Mungu reservoir located downstream.

There are many large scale users including the Tanzania Electric Supply Company (TANESCO), estates (coffee, horticulture and flower companies), and cities. TANESCO owns five hydropower facilities on the Pangani River, contributing about 17% of electricity to the national grid.

The spiralling water demand is a source of competition and conflict between users within and outside the catchment. Tensions and sometimes violent confrontations occur between smallholder farmers and cities within the catchment (Komakech et al. 2012e), and between large commercial farmers holding government water use permits and smallholder farmers relying on customary water access rights (Komakech et al. 2012a). Every year TANESCO attributes drops in its power production to wasteful water use by smallholder farmers. So far attempts by the basin water board to regulate water use through issuing water permits and construction of diversion gates have not solved water allocation conflicts (Komakech et al. 2011b). Many of the diversion gates constructed between 1994 and 1997 have been vandalised. Recently smallholder irrigators have started using mobile water pumps making it even more difficult for PBWO to regulate water use. Water pollution from the two fast growing cities is also increasing. These challenges and the government policy of decentralisation by devolution of management responsibilities provided the backdrop for the PBWO to introduce catchment forums.

4.4.3 Kikuletwa catchment institutional environment and actors

The Kikuletwa institutional environment is a mosaic of locally evolved arrangements and state-led and NGO-created forms of water management. In addition, a diversity of actors has interests in the water resources of the Kikuletwa catchment. We categorised the existing institutional arrangements as state-led or locally evolved. However our description is far from complete and should be seen as an attempt to sketch a complex situation.

State-led water institutional arrangements

Tanzania´s water policy and associated legislation provides for the establishment of formal catchment and sub-catchment water committees, and water users associations (Tanzania 2002a, Tanzania 2009). The committees and water users associations are meant to coordinate and harmonize integrated water resources management plans, resolve water conflicts, and perform other delegated functions. A water users association (WUA) may be formed by agreement of the majority of users of a common stream with the aim to: allocate water, acquire a water use permit, resolve water conflicts between its members, and collect water use fees on behalf of the basin water board (Tanzania 2009).

In the past the PBWO has created some WUAs, but very few remain functional (Box 4.1). According to Pamoja (2006), most of the WUAs created were single purpose user associations of irrigators. Some of the WUAs were registered as cooperative societies with the Ministry of Agriculture, Food Security and Cooperatives, while others were registered with the Ministry of Home Affairs as associations. The unclear registration process has made some of the registered users to consider the water user association not "legitimate enough" to sanction their claims for water.

Box 4.1: Tegemeo water users association (Source: adapted from Pamoja, 2006).

Tegemeo operates in Rundugai ward (Hai District). The Association covers five villages namely Rundugai, Kawaya, Mkalama Chekimaji and Chemka and represents about 900 households. It is reported that initial attempts by Tegemeo WUA to collect water user fees led to a misunderstanding between the management of the WUA and the village governments. This is because in the past each village had its own way of collecting revenue. The village governments felt the WUA leaders do not have authority to collect the water user fee. Attempts to prepare a seasonal calendar showing types of crops to be produced in which season of the year also failed. In the cropping calendar, paddy cultivation was prohibited in the dry season. Paddy rice farmers rejected the calendar because dry season paddy fetches high market prices. The issue was brought to the government court but was never resolved.

Local self-organised institutional arrangements

Alongside the state forms of water governance co-exist the locally evolved water sharing arrangements that draw on local norms, customs and traditions. Many irrigation canals (locally called furrows) are managed by smallholder farmers (Komakech et al. 2011b). Furrow leaders regulate water access for different users and arrange for periodic maintenance. In many cases they constitute the main link between the farmers and the state-led WUA and the village government. Individual farmers' access to irrigation water from a particular canal is based on access to land in the command area, provision of labour for maintenance, affiliation to social networks and, in some cases, payment of entrance fees (Gillingham 1999). Some of the furrows share river intakes and head canals and have formed water user groups to manage water allocation between them (Box 4.2).

Box 4.2: Olbuso water users association (Source: adapted from Pamoja, 2006).

Olbuso water user group comprises three villages (Shambarai Burka, Shambarai Sokoni and Olbili) that share Olbuso main furrow. The furrow serves an estimated 9,000 people, roughly about 3,000 per village. The group applied for a water right in 1997 and was granted a collective right of 200 l/s by Pangani Basin Water Office. Each village is represented in the water users association by its village water committee, village chairman and village executive officer. In total there are 75 representatives. Every three years, the 75 members elect a new management committee - General Manager, Deputy General Manager, Treasurer, and Secretary. All the three village chairmen and village executive officers are also members of the management committee.

Olbuso water user group is responsible for water allocation to the villages, arranging for maintenance, conflict resolution, and payment of a collective water user fee to PBWO and representing the interest of the three villages at the Kikuletwa river committee level. They meet once a week during the dry season (normally on Thursday). The water users contribute money for canal maintenance, water user fee, and allowances for the general manager to attend the river committee meetings. The users' contributions are collected by the village water committees.

Water conflicts are solved in a graduated manner. It is first tried by the water distributor of each village furrow, then by the village furrow water advisors (normally elders); if they fail the case is forwarded to the chairman of the village furrow, then the village water committee, the village chairman, the water user group and eventually to the river committee. If they all fail to resolve the conflict it is either forwarded to the Division Secretary, the District Council or to the PBWO.

Some furrows have federated into a WUA and then registered as a cooperative society with the Ministry of Agriculture, Food Security and Cooperatives (Box 4.3). The registration allows them to access loans from banks, apply for collective water rights and operate as an institution for credit and saving (Pamoja 2006).

Box 4.3: Mbukita water users association (Source: adapted from Pamoja, 2006).

Mbukita is an association of the three villages of Mbuguni, Kikuletwa and Msitu wa Mbogo served by Kikuletwa, Msitu wa mbogo and Kambi ya tanga mama furrows. The main intake supplying the three furrows is at Kambi ya tanga and is referred to as Mbukita furrow. The Association was first established in 1997 and was registered as a cooperative society in 2001 with the Ministry of Agriculture, Food Security and Cooperatives. The farmers applied for a water use right in 1997 and were granted a collective water right of 200 l/s by PBWO. The association is managed by an elected board of nine members. Under the board are three committees namely: 1) finance and planning responsible for accounting and development planning; 2) construction committee responsible for maintenance; and 3) irrigation management committee responsible for water allocation and conflict management.

The board's primary responsibility is water allocation, conflict resolution and payment of water user fees. All members must pay a one off registration fee of Tsh. 200; buy five shares each worth Tsh. 5000; and pay an annual membership fee of Tsh. 1000. In total there are 1000 users but only about 300 users have registered with the association. An elaborate procedure has been put in place for members and non-members of the association to access water as follows: non- members in Mbuguni ward using water must pay an irrigation season fee of Tsh. 37,000 per hectare. Non-members who live in other wards but farm in Mbuguni pay an irrigation season fee of Tsh. 120,000 per hectare. The association members who have rented land outside Mbuguni ward but use water pay an irrigation season fee of Tsh. 12,000 per hectare. Members of the association pay a water distribution fee of Tsh. 500 per irrigation season. Water theft is fined Tsh. 50,000. In addition, the association represents the interest of its member at the Kikuletwa river committee. The WUA works through the river committee for conflicts with other users of the Kikuletwa river. The Mbukita board is represented by the chairperson and vice chairperson in the Kikuletwa river committee. In the Kikuletwa river committee, the Mbukita chairperson was the general secretary during the period of field work.

River committees have emerged to manage water allocation and resolve conflict between groups of users, both large and small, using the same river source (Komakech and Van der Zaag 2011). In total, seven river committees have been identified in the catchment (four in Themi, and one each in Nduruma, Weruweru and Kikuletwa). Most river committees in the catchment operate independently and do not presently communicate with each other (Komakech and Van der Zaag 2011). The committees also do not formally interact with the basin water board/office; however, they do work with local government institutions, i.e. district departments, and ward and village offices. The local government institutions consider these self-initiated river committees legitimate and valuable in the local water management hierarchy.

4.5   PROCESS AND FORMULATION OF KIKULETWA WATER USERS

ASSOCIATIONS

In 2003, the International Union for Conservation of Nature (IUCN, an international NGO), PAMOJA and PBWO entered into a partnership to implement a so-called dialogues project. This project, under IUCN's Water and Nature Initiative (WANI), sought to mainstream the ecosystem approach in catchment and river basin management. In the Pangani it sought to contribute to the efficient water resource management by building local capacity to negotiate equitable solutions to water conflicts. A number of pilot activities in five sites in the Pangani river basin were carried out between 2003 and 2004. They included irrigation infrastructure improvement, creating dialogue platforms and facilitating negotiated agreements between local water users. Through a basin situational analysis study, several key challenges for water management were identified. One of the most significant of these challenges affecting water allocation in the basin was the rapidly increasing water demand due to population growth and economic activities. Based on the experiences gained, the partners initiated a new project to improve water governance in the Pangani river basin using the concept of integrated water resources management. A component was the establishment of catchment and sub-catchment forums. It was argued that water rights allocation to individual users would be better debated and resolved at the catchment and sub-catchment level forums. Other issues, such as releasing water from an upper catchment to a lower catchment (e.g. to meet downstream needs related to hydropower and environmental flow requirements) were considered to be best analyzed and debated at the basin level.

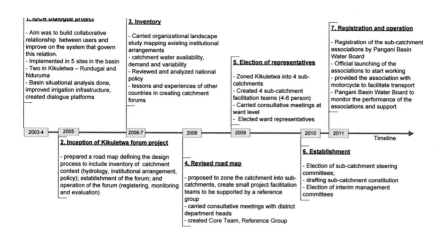

Figure 4.2: Process timeline for the establishment of Kikuletwa catchment forum.

PBWO and the development partners then embarked on designing catchment forums and the Kikuletwa catchment was selected as a pilot case. In 2005 a road map for the

design of the Kikuletwa catchment forum was developed and SNV, a Dutch development organisation, was contracted to coordinate the forum process.
Figure 4.2 presents the timeline of the Kikuletwa catchment forum project.

However, implementation of the project experienced delays. The inventory phase was only completed in 2007 and things stalled thereafter for nearly a year. Overall, the project turned out to be more complex than anticipated by the partners. First, there was lack of a common understanding what the forum was supposed to be. A complicating matter was that the catchment forum concept was not explicitly mentioned in the National Water Policy (2002). The National Water Sector Development Strategy 2006-2015 and the Water Act of 2009 only made mention of catchment and sub-catchment committees, being autonomous bodies financed from user charges that can be created to carry out functions delegated by the basin water board (Tanzania 2002a). The committees were foreseen as modest offices with a small number of part time staff and with minimum administrative expenses. In contrast, IUCN conceived a catchment forum much more ambitiously, as an arena where competing actors or their representatives can meet and dialogue on conflicting issues and find common ground. As a result, the partnership got locked in theoretical discussions of what a catchment forum was supposed to be and how it should be established.

Second, the organisational landscape study carried out by Pamoja (2006) identified several user groups and institutions active in the catchment. How these actors and institutional arrangements would be involved in the process remained unclear. The project partners perceived the forum as something that would first be designed and then subsequently be given to the water users to implement. Third, the large size of the catchment further complicated this design.

In 2008, the forum process gained momentum again as the road map was revised. In the new approach, a two-person project core team was created to run the process and this team was also made responsible for the project output. A reference group was constituted to guide the overall process. Recognising the large size of the catchment it was decided to divide it into sub-catchments and the focus turned to creating sub-catchment forums in each of them. It was envisaged that the sub-catchment forums would later federate to form the Kikuletwa catchment forum. In 2009, Kikuletwa was zoned into four sub-catchments as follows: 1) Upper Kikuletwa, 2) Sanya–Kware, 3) Kikafu-Weruweru-Karanga, and 4) Lower Kikuletwa sub-catchment (Figure 4.3).

However, in 2010 the sub-catchment forums were renamed sub-catchment WUAs. First, this was decided after the realisation that in the Water Act of 2009 catchment or sub-catchment areas were to be declared by the order of the Minister of Water, which would entail a cumbersome administrative procedure. Second, the Water Act (2009) envisaged catchment and sub-catchment committees to be small entities with 3-5 members including the chairman, with at least one representative of major private sector water users, up to two representatives of existing WUAs, and one from the local government authorities in the catchment area (Tanzania 2009). Third, it provides that all catchment and sub-catchment committee members except the local government representative are to be appointed by the basin water board. Given the large number of different types of users (smallholder, commercial farmers, cities etc) and districts in the catchment, it was nearly impossible to come up with meaningful representation. Thus the phenomenon of "sub-catchment WUA" was created.

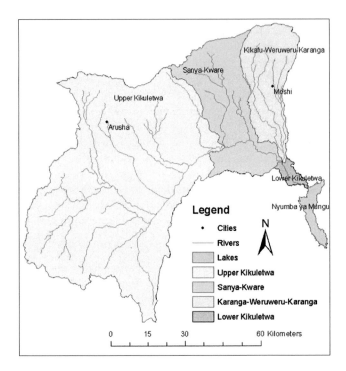

Figure 4.3: River systems under the Kikuletwa sub-catchment water user association.

In creating the four sub-catchment WUAs, representatives were selected from each of the fifteen tributaries of Kikuletwa River. They were selected from elected ward representatives during several stakeholder consultative meetings conducted by the core team and four sub-catchment facilitation teams. In these meetings, each ward elected about 10 representatives (over seventy wards elected representatives). Through four training workshops, the ward representatives were facilitated to elect from among themselves the representatives for each of the fifteen rivers to form the four sub-catchment WUA committees.

Figure 4.4 shows the overlap and compromise between hydraulic and administrative entities; it depicts a complexity which is often overlooked in the discourse on catchment management and shows how in the Kikuletwa this was attempted to be resolved.

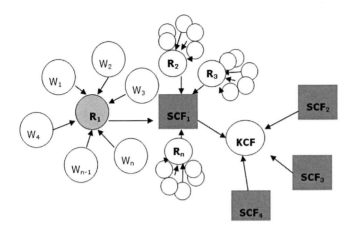

Figure 4.4: Schematic representation of the sub-catchment WUA committee selection process. W1-n are the elected ward representatives, R1-n are the elected representatives for the river systems, SCF1-4 are sub-catchment WUA committees, and KCF is the Kikuletwa catchment forum to be formed by the four sub-catchment WUAs at a later stage.

After the four WUAs committees were established, their members were trained to draft a constitution of their respective sub-catchment WUAs. The WUA constitutions, completed in August 2010, detailed their institutional structure, roles and functions (Figure 4.5).

The supreme body of each sub-catchment WUA is the general assembly of all river committees in the sub-catchment. The registrar role, performed by the Pangani Water Board, includes registration of the association and technical support related to water resources planning and conflict management. The sub-catchment WUAs are expected to have offices and to employ a small number of staff to manage the association records.

Attempts were made to integrate local arrangements; as such the local river committee was included in the WUA structure. However, the river committees mentioned in the sub-catchment constitutions are not the existing river committees created by the water users (Komakech and Van der Zaag 2011). River committees were thus created afresh. In Themi River, for example, a new river committee was created. Some members of the existing Lower Themi, Seliani, Burka and Ngarenaro river committees were seconded as representatives to the new Themi river committee. In Nduruma also a new river committee was formed, being a federation of the upstream and downstream water committee. The upstream committee was newly created, while the downstream was the existing Nduruma river committee originally created by the mid- and lowland farmers (See Komakech et al. 2012a). The sub-catchment WUAs were subsequently formally registered by the basin water board and inaugurated in early 2011. The Kikuletwa apex catchment forum had not been established by 2011. PBWO stated that the apex forum would be created at a later stage when the sub-catchment WUAs are in full operation. The WUAs have been provided with office space and two motorcycles each and are encouraged to start registering water users in their areas of jurisdiction, implement water source protection laws, and resolve water conflicts.

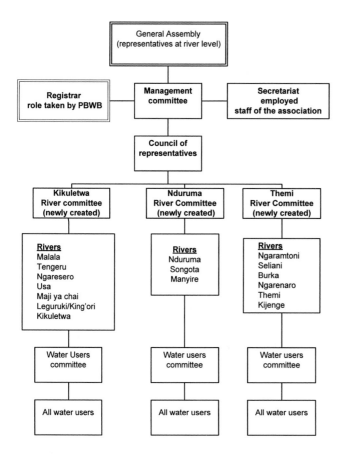

Figure 4.5: Proposed institutional arrangement of Upper Kikuletwa sub-catchment WUA (Source: PBWO, 2010).

However, the WUAs have encountered difficulties in exercising their authority over existing local arrangements. The authority of Sanya-Kware sub-catchment WUA, for example, has been questioned by local resource users in the area, in particular those who rely on the Boloti wetland (Box 4.4).

Box 4.4: Sanya-Kware WUA struggles to gain control of Boloti wetland (Source: PBWO).

Boloti wetland is located within Sanya-Kware sub-catchment. The wetland receives water from the small Kishenge river and its outlet drains into Sanya river. It is being used by Bondeni Estate and two villages, namely Munguishi and Kyuu of West Masama and South Masama wards respectively. Irrigation water is extracted from the wetland using pumps and buckets. The farmers mainly grow coffee, maize, banana, yams, tomatoes and vegetables. Both Bondeni estate and the two villages have encroached into the wetland. In April 2010 the villagers formed an environmental group called Green Guard to manage and protect the wetland. They planted 2000 trees to demarcate the wetland area.

In March 2011, the chairman and secretary of the newly created Sanya-Kware WUA went to Boloti and ordered farmers to vacate the wetland area. They also told farmers that all water users must register with the WUA and should pay a membership fee of Tsh. 20,000 to the WUA. This angered the Kyuu villagers and members of the Green Guard, who petitioned the Environment Secretary of Masama West Ward. The Environment Secretary wrote a letter to the Pangani Basin Water Officer informing him that members of Sanya-Kware WUA wanted to own Boloti wetland. He said the WUA was not known to the villages. He also explained that the WUA represented only few people in Kware and Sanya area and none from the villages using Boloti. When a staff member of PBWO went to confirm that the WUA was indeed responsible for water management in Sanya-Kware sub-catchment, including Boloti wetland, the villagers again challenged the WUA's authority and legitimacy. The matter was taken to the government police by the WUA for settlement. After consultation with the villages, PBWO and WUA, the police advised the parties to solve the conflict outside the court.

PBWO later organised a meeting with the villagers and leaders of Sanya-Kware WUA. In the meetings PBWO informed the villagers that the WUA was created following consultations with Hai district and wards within Sanya-Kware between September 2009 and March 2010. PBWO cites the Environment Management Act of 2004 and the Forest Act of 2002, stipulating both that wetlands were water sources that needed to be protected, and further, that according to the Water Resources Management Act of 2009 all water users must acquire a water use permit. The villagers said they had never heard of any meetings or elections of the WUA. They see the WUA as an association representing only few people and noted that the awareness campaign on water resources management carried out by PBWO reached only few people and even those who attended the workshops never provided feedback to the rest of the villagers. The villagers also claim that they have never seen the draft constitution of the WUA.

The Upper Kikuletwa sub-catchment WUA started with registration of water users within its area of jurisdiction, but progress was hampered by the large area of the sub-catchment. The existence of the WUA remained unknown to most of the water users. The Lower Themi furrow committees report that they are not aware of the existence of this WUA and its role. Although some members of the Lower Themi river committee have attended the Upper Kikuletwa sub-catchment WUA meetings, they have not provided feedback to the furrow committees. The Kikafu-Karanga-Weruweru and Lower Kikuletwa sub-catchment WUAs have not functioned since their establishment despite being given motorcycles. By the time of concluding field work they were yet to start registering their members.

All four sub-catchment WUAs are facing financial difficulties. The sub-catchment WUAs were envisaged to be financed through their members' registration fees, annual contributions and fines. According to the PBWO, the water users should finance the operations of the sub-catchment WUAs since the associations were created in their interest, while it would provide technical support. However, none of the WUAs have been able to collect such moneys yet. Currently all the sub-catchment WUA leaders self-finance their operational costs. This has affected their operation. The motorcycles have run out of fuel and have been parked.

4.6 DISCUSSION: WATER INSTITUTIONAL DESIGN PITFALLS

There is a general belief that users' participation in dialogue and decision making over water allocation and management will reduce conflict in a catchment (Jaspers 2003). But this requires creating an institutional structure that has legitimacy and is recognised as such by all relevant actors. The catchment forum concept may be a good idea for effective participatory management of water stressed basins. However, its implementation in the Kikuletwa catchment faced many problems. In this section we highlight some of the challenges.

First, Kikuletwa is a large and complex catchment spanning six districts and a total of fifteen tributary rivers. Some of the rivers are dry for most part of the year due to overuse. It is therefore difficult to define the most appropriate hydrological management unit for decision-making, especially if it also has to fit with the political-administrative territories.

Second, the actors with interests in Kikuletwa are diverse and extend far beyond its hydrological boundaries to include international NGOs, development banks and governments. Selecting representatives proved difficult. Large users in the catchment such as large commercial farmers and TANESCO never participated in the forum. The hydropower stations of TANESCO are located downstream of the catchment but the parastatal company is able to influence decisions at the basin water board.

Third, the institutional arrangements in the Kikuletwa catchment are messy with overlapping jurisdictions between state-led and locally created institutions. The national government has attempted to restructure the spaces for participation through establishing the basin water board, catchment and sub-catchment committees and WUAs. However, these state-led arrangements are being layered on top of pre-existing local institutional arrangements. At the level of tributaries of the Kikuletwa river, water users have organised themselves into furrow committees, WUAs and river committees. Furrow committees work closely with the local village governments to allocate water and manage conflicts among individual farmers. Where two or more furrows have formed a joint WUA to manage water allocation between furrows sharing one river intake, the association is often registered as a cooperative society to secure loans from banks. River committees manage the allocation of water between the users of a part of a river and appear to be able to solve the coordination challenge experienced by upstream and downstream, large-scale and small-scale farmers (Komakech and Van der Zaag 2011, Komakech et al. 2012a). The success of these locally created water institutions is because they are considered legitimate by the water users and the local government institutions.

Recognising the complex environment in Kikuletwa, the Pangani Basin Water Board and partners chose first to form four sub-catchment WUAs and subsequently to create the apex organisation. The apex organisation has so far not been established. It can be argued that the modularisation of Kikuletwa into sub-catchment WUAs allows polycentric governance that nests local arrangements. In practice, however, the new Kikuletwa sub-catchment WUAs are like islands of associations not well integrated with the existing arrangements. Water users do not see how the WUA is linked to their own governance arrangements (e.g. furrow and river committees) and constantly ask "how do we benefit from paying memberships and annual fees to the WUA?" This is not surprising because the process of forming the sub-catchment WUAs was highly centralised and can at best be described as a top-down approach that was branded as bottom-up. The forums were designed in the office, rolled out from the centre and later handed over to the users. Although an organisational landscape study identified several local institutions (Pamoja 2006), the forum designers were more interested in designing new structures.

The sub-catchment WUAs were envisaged to improve the active representation of water users in water management. As highlighted by the struggle over control of the Boloti wetland, the local farmers feel that the new associations benefit a small group of users only - in fact Boloti wetland farmers indicated that they are not willing to pay fees to an association that does not, in their view, have a mandate in water management. PBWO and development partners, however, maintain that the sub-catchment WUAs do not diminish the legitimacy of locally established institutions. They argue that WUAs build upon these arrangements whilst allowing the local institutions to continue to function at the lower level (e.g. furrow level), but that the sub-catchment WUA committee members were not sufficiently trained how to engage with existing local governance structures.

In addition, the sub-catchment WUAs may provide large water users an opportunity to strengthen their power network and sustain inequitable water access and control. We observed this power dynamic in the struggle over the control of Boloti wetland. The current owner of Bondeni estate claims that the wetland is part of his coffee estate and in the past he has tried to evict farmers but failed. The estate manager now supports the chairman of Sanya-Kware sub-catchment WUA. According to the estate manager the farmers are encroaching on the wetland thereby destroying its ecosystem and should be evicted. However, interviewed farmers say they were allowed to settle around the wetland by the first estate owner in the 1970s and that they have since made significant investments in the land.

PBWO and its development partners appear to have missed the opportunity to upscale locally evolved institutional arrangements. The locally created river committees could have been entry points for engaging meaningfully with the water users. The Kikuletwa project could have strengthened the river committees as forums where most if not all river users are represented. Currently most of the existing river committees only govern sections of rivers (see: Komakech and Van der Zaag 2011, Komakech et al. 2012a). But the river committees' objectives align well with ideas underlying the concept of integrated water resources management, for example relating to managing water along hydrological boundaries and involving users in decision-making. The river committees also integrate water management with local government institutions (villages, wards and districts). They are therefore able to transcend the problem of administrative boundaries and institutional fit (Young 2003, Ekstrom and Young 2009). The river committees could therefore become sub-units (cf. Lankford and Hepworth 2010) in the catchment. PBWO could see to it that the sub-units commit to transferring certain minimum flows to downstream. However, this requires a sound and detailed knowledge of local water resources. In the Kikuletwa catchment, PBWO and partners tried to resolve the problem of administrative boundaries and institutional fit by selecting users' representatives at the river level from each of the administrative wards comprising a particular river. But this was insufficient to integrate customary arrangements with the state-led governance structure.

4.7 CONCLUSIONS

Active participation by the water actors is often considered to lead to better decision making and coordination. To engage actors in water dialogues catchment forums have been proposed. Using the case of Kikuletwa, this paper highlights the difficulties of designing catchment forums in an African context. Creating a catchment forum in Kikuletwa catchment was challenged by its large spatial coverage and its complex river system. The diversity of actors with interests in the water resources made it difficult to find an appropriate representation model for water users. The multiplicity of institutional arrangements found in Kikuletwa complicated the problem of administrative and institutional fit.

We explored the relevance of a polycentric governance approach as a framework for integrating local and state institutions.

The Pangani Basin Water Board and partners created four sub-catchment forums (in the shape of WUAs) that would later federate to form an apex Kikuletwa catchment forum. In so doing they tried to nest and upscale institutional arrangements. But the sub-catchment WUAs were weakly linked to existing institutional arrangements, which made them ineffective.

An alternative strategy to promoting effective dialogue forums is to creatively strengthen local water management practices and organisations (Warner et al. 2008, Merrey 2009, Merrey and Cook 2012). This approach is based on the idea that institutions evolve through bricolage - a complex creative process where multi-identity actors adopt and adapt collective action mechanisms from diverse sources including existing rules, norms, styles of thinking, social relationships and social identities (Cleaver 2002, Merrey and Cook 2012). To succeed, any government, development organisation or agency planning to create dialogue forums would need to understand the local water resource management context; that is they should first invest in research to identify existing arrangements and understand their strength and limitations. Subsequently, and based on this understanding, a program can be developed to strengthen the positive aspects of the existing institutional arrangements while attempting to minimise some of the negative aspects such as gender inequity, power relations over water and control).

In the Kikuletwa, a pragmatic starting point for encouraging meaningful dialogue would have been to build on existing river committees in each of the fifteen major tributary rivers that comprise the catchment. Komakech and van der Zaag (2011) discuss the emergence of three river committees in a sub-catchment of Kikuletwa. The river committees were all crafted using the existing principle of good neighbourliness, the rationale of local water allocation (e.g. an innovative, transparent and locally developed system of water allocation that is perceived as proportional and therefore equitable), and a traditional system of conflict resolution (e.g. the age group system to guard and manage water adapted from the Maasai). The biggest challenge for this pragmatic approach would be to overcome one fundamental weakness of the water committees, namely that most of the existing river committees only manage parts of a river. Finding a way to motivate distant upstream users to agree to water sharing arrangements would be key. The river committees would continue to manage water allocation between users within their river reach and could be issued a collective water use right with a condition to ensure some minimum outflow during the dry season for downstream use. This is already happening in other parts of the Pangani basin (see Box 4.2 and Box 4.3). The Pangani Basin Water Board could then concentrate its efforts on monitoring the outflow from each river, and penalise committees if the minimum flow conditions were violated.

The Kikuletwa catchment forum process highlights the difficulties of crafting institutional arrangements that can coordinate activities at catchment and basin scales. The paper demonstrates the need of linking such larger scale initiatives with existing locally evolved arrangements. Resolving the problem of institutional fit while integrating customary arrangements with the state-led governance structure requires careful analysis of local structures.

Chapter 5

The Last will be first: Water transfers

from agriculture to cities in the Pangani

river basin, Tanzania[7]

5.1 Abstract

Water transfers to growing cities in sub-Sahara Africa, as elsewhere, seem inevitable. But absolute water entitlements in basins with variable supply may seriously affect many water users in times of water scarcity. This chapter is based on research conducted in the Pangani river basin, Tanzania. Using a framework drawing from a theory of water right administration and transfer, the chapter describes and analyses the appropriation of water from smallholder irrigators by cities. Here, farmers have over time created flexible allocation rules that are negotiated on a seasonal basis. More recently the basin water authority has been issuing formal water use rights that are based on average water availability. But actual flows are more often than not less than average. The issuing of state-based water use rights has been motivated on grounds of achieving economic efficiency and social equity. The emerging water conflicts between farmers and cities described in this chapter have been driven by the fact that domestic use by city residents has, by law, priority over other types of use. The two cities described in this chapter take the lion's share of the available water

---

[7] Based on Komakech et al. 2012. *Water Alternatives* 5(3), 700-720

during the low-flow season, and at times over and above the permitted amounts, creating extreme water stress among the farmers. Rural communities try to defend their prior use claims through involving local leaders, prominent politicians and district and regional commissioners. Power inequality between the different actors (city authorities, basin water office, and smallholder farmers) played a critical role in the reallocation and hence the dynamics of water conflict. The chapter proposes proportional allocation, whereby permitted abstractions are reduced in proportion to the expected shortfall in river flow, as an alternative by which limited water resources can be fairly allocated. The exact amounts (quantity or duration of use) by which individual user allocations are reduced would be negotiated by the users at the river level.

5.2 INTRODUCTION

Urban centres are steadily growing and need more and more water. Transferring water from agriculture to cities is an obvious way of reallocating the uses and users of the available water in a catchment (Celio et al. 2010). The main rationale is that in situations of water scarcity allocation should favour uses with the highest returns per unit of water (including basic human needs). In this discourse, agriculture is considered as a voracious user of water that mainly produces low-value output (Savenije and Van der Zaag 2002, Molle and Berkoff 2006). In cases where the level of water development has reached its maximum and/or inter-sectoral reallocations do not suffice, interbasin transfers seem to be the preferred (supply-oriented) strategy (see Swyngedouw 1997 on urbanisation of water, for an assessment of interbasin water transfers see Gupta and Van der Zaag 2008). Little attention is being paid to the fact that irrigation water may serve many other high-values uses (e.g. domestic, vegetable gardens, livestock, fishing, and construction).

In Tanzania, the ongoing state-led formalisation of water allocation may be considered a continuation of a process started by the British colonial power. As early as 1923, the British colonial administrators introduced a statutory water right system in mainland Tanzania (then Tanganyika), whereby the ownership of all water resources was vested in the King of England. Water rights were issued to users located in areas declared crown land, while areas under "natives" were allowed to be governed by local customs and traditions.

Although the independent government of Tanzania was at first preoccupied with modernisation through irrigation development and the reorganisation of villages, it subsequently amended the colonial water law and policies, introducing water rights fees and volumetric charges for water used. The most recent attempt by the government to regulate water use is driven by increased scarcity, which among others is manifested by frequent electricity power cuts (Lankford et al. 2009). Nearly all Tanzania hydroelectric power plants are located downstream of other users and are, hence, very sensitive to water scarcity. Water shortages are attributed to uncoordinated planning of use, imperfect policies, inefficient use in the agricultural areas, and inadequate monitoring (World Bank 1996). To solve this problem the water policies and laws were revised in 2002 and 2009, respectively. The National Water Policy of 2002 gives first priority to water for basic human needs (often interpreted as water for drinking only, not considering other domestic needs), second priority is given to water required to protect ecosystems, while all other uses are subject to social and economic criteria to be reviewed from time to time (Tanzania 2002a). The policy recognises that "water is a public good of high value in all its competing uses, and requires careful conservation and sustainable utilization" (Tanzania 2002a). It cites extensive irrigation during dry seasons and inefficiencies of many irrigation schemes as major causes for reduction in water availability (Tanzania 2002a). This position is in line with generally held views that: (a) agriculture gets the lion's share of all water diverted and yet generates low returns per unit water used; (b) agriculture incurs the largest wastage; (c) water productivity in the non-agriculture sector is much higher than in agriculture; and (d) cities are frequently water-short (Molle and Berkoff 2009, Rosegrant et al. 2009). Thus it is believed that considerable gains can be achieved by improving irrigation efficiencies and, if that is not sufficient, through reallocating water to higher-value uses (Molle and Berkoff 2009). In the Pangani basin irrigated agriculture is mainly practised by smallholder farmers. It is believed that these farmers utilise most of the available water but with very low efficiencies leading to water stress (Maganga et al. 2002, Kashaigili et al. 2003).

Implementation of the 2002 water policy, however, appears to generate conflicts in water allocation at the local level. Thirsty cities within a river basin refer to the water act, which gives priority to domestic needs, to claim water already in use by rural communities for small-scale irrigated agriculture. This leads to tensions and sometimes violent confrontations. This chapter describes and analyses processes of water appropriation from smallholder irrigators by cities in the Pangani river basin and the ensuing conflicts. Using a framework of agriculture-to-city water transfers, it identifies shortcomings in the current water allocation system and proposes an alternative allocation mechanism that takes into account the variability in supply and also proposes alternative institutional arrangements for its enforcement.

The following section (5.3) reviews the concept of water allocation focusing on water right administration and transfer and highlights the typology of transfer and mechanisms often used. Section 5.4 introduces the study area (Pangani river basin and study sites) and the research methods used. Section 5.5 presents two cases of city versus smallholder farmers' water allocation conflict. The next section (5.6) discusses the findings and by way of conclusion (section 5.7) the chapter explores mechanisms by which limited water resources can be fairly allocated between cities and rural areas.

5.3 CONCEPTUAL REVIEW: WATER TRANSFER BETWEEN

AGRICULTURAL AND URBAN USE

When water is scarce it has to be shared among competing interests and this requires putting in place criteria and procedures that clearly define who is entitled to what amount of water, at what time, for how long and in which place. In addition, proper institutional arrangements with means to monitor the enforcement of the water-related rules are required. Although the arrangement can also be developed by users, religious communities, non-governmental organisations and customary leaders, normally governments assume the role of the main regulator of water use in a catchment. State-led water management reforms have included the formalisation of water right[8] administration and the creation of basin management institutions. Water ownership is vested in the state and users are required to acquire permits to use water from a given source. These approaches are used to justify government's intervention in water allocation in terms of economic efficiency, social equity and sustainability of the water resource (Syme et al. 1999, Wang et al. 2003). These three principles, coupled with the notion of users' participation in the decision-making process, are integral components of the discourse on Integrated Water Resources Management (IWRM).

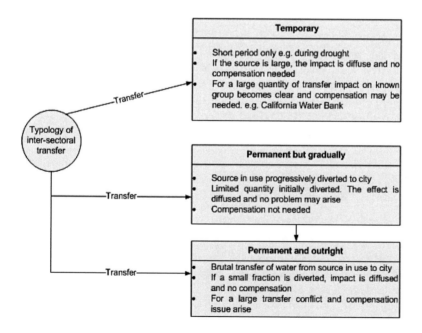

Figure 5.1: Typology of water transfer (Source: adapted from Molle and Berkoff 2006, Meinzen-Dick and Ringler 2008).

---

[8] In this chapter, a water use right and a water permit have the same meaning – both confer a time-bound right to beneficial use of the available water but not its ownership.

In places where water is over-allocated, reallocation and transfers between uses or sectors are the typical responses (other responses may include reuse of treated wastewater, improvement of irrigation efficiency, etc.). The process of water transfers from agriculture to cities takes several forms and may include: temporary transfer; permanent but gradual transfer; and permanent and outright transfer (Figure 5.1). Temporary transfers typically occur during periods of drought; the agriculture sector will be severely affected albeit for a limited period of time. The most quoted example of temporary water transfer is the California Drought Water Bank which arranges temporary water purchases from individual farmers for transfer to other users (Molle and Berkoff 2006, Meinzen-Dick and Ringler 2008). Permanent but gradual transfer is a case where a water source is progressively diverted to the city. Initially a limited quantity may be diverted, which diffuses its effect in the short term. Permanent and outright transfer, on the other hand, is a sudden and long-term reallocation; large transfers are often contested by the existing users.

To understand the impacts of transfer it is interesting to follow the mechanism (formal or informal) by which the three types of water transfer are implemented. Four mechanisms of transfers can be identified (Meinzen-Dick and Ringler 2008, Molle and Wester 2009). First, *market-based mechanisms* allow water to be sold either directly to buyers for non-agricultural uses or indirectly through transfers of land with a water right appurtenant to it. Second, *water right transfer through administrative decisions* follows a formal procedure which is spearheaded by a national government or basin management institution according to the functions assigned by law. Celio et al. (2010) highlight how water transfers from the Krishna and Manjira rivers to the city of Hyderabad in India were sanctioned through several government orders. Prior use rights are rarely recognised although indirect compensation may be given (Meinzen-Dick and Ringler 2008). Farmers may protest against administrative transfers but are mostly unsuccessful against cities that appear to be more powerful. Third, *transfer through collective negotiations* with communities can be concluded between existing users and the state or between the users themselves. Collective negotiations aim at win-win solutions and may take into account other uses (Molle 2004). Fourth, *transfer by stealth* is done unilaterally by the state, basin authority or other entity, without complying with formal procedures and/or legal requirements, and without consulting those potentially affected.

Although widely promoted, available literature on water transfers indicates that they often have negative impacts on irrigators, other uses linked to irrigation water, and the environment (Hearne 2007, Molle and Wester 2009, Celio et al. 2010, Movik 2012, Perramond 2012). Meinzen-Dick and Ringler (2008) present a case where water-exporting regions in California lost more in crop production than they were paid for the water. Market-based transfers, e.g. water sales by tankers' association from rural to middle-class residents, have been reported (Meinzen-Dick and Ringler 2008), but water markets at larger spatial scales have been less frequent and often unsuccessful, partly because of the infrastructure needed to transfer water from one user to another (Molle and Berkoff 2009). So far, positive experiences of market-based transfers are confined to countries with strong legal, institutional and regulatory backgrounds and relatively wealthy stakeholders (Hearne 2007, Molle and Berkoff 2009). Thus in countries characterised by data scarcity, where the requisite physical infrastructure for water control (e.g. storage reservoirs, canals) is lacking, and with weak monitoring and enforcement capacity, water transfers by market-based mechanisms are likely to be problematic.

Finally, in most African countries state law is not the only source of water rights but there are also customary rules backed by local authority and social norms that govern water access. Religious laws and development projects also define the condition for access to water. Hence, users may use different rules and rights to claim water access (Meinzen-Dick and Pradhan 2002). The plurality of water laws may be a source of conflict when dealing with water reallocation.

5.4   STUDY AREA AND RESEARCH METHODS

Study area: Location of research catchment within the Pangani river basin

This chapter focuses on the Kikuletwa catchment which is located in the upper part of the Pangani river basin (Figure 5.2). The study focuses on the water struggles between the city of Arusha and Moshi and the surrounding villages.

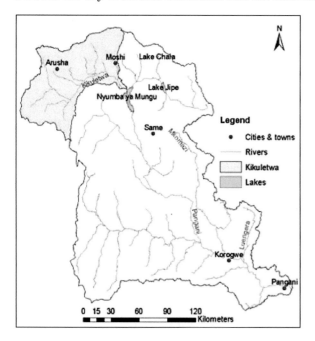

Figure 5.2: Pangani river basin, reservoir, lakes, cities, towns and Kikuletwa the case study catchment.

The two cities of Arusha and Moshi are located in the upper part of the basin (Figure 5.2). The development in these urban areas is partly driven by their location in regions with productive agriculture, mining activities and a booming tourism industry. In 1977, both cities had about 50,000 residents each. In 2010 number of residents of Moshi and Arusha had increased to 156,000 and 367,000, respectively. This growth puts significant pressure on the basin's limited water resources in terms of water for domestic, commercial and industrial uses and crop production.

The fast expansion of irrigated areas and increased cropping intensities, rapid urbanisation and an increased demand for water from cities, combined with climate variability have resulted in many tributaries of the Pangani river now only flowing in parts of the year, i.e. during the rainy seasons (Mul 2009). The basin is therefore experiencing stiff competition and conflict over its water resources. Conflicts between city water authorities and smallholder irrigated agriculture; between farmers and hydropower facilities located downstream; and between large commercial farmers and small-scale irrigators are all increasing in both scale and frequency (Komakech et al. 2011b).

Research material and methods

To understand the implication of urban water appropriation on smallholder farmers we studied the historical processes through which water transfers had taken place. Field studies were conducted (January to March 2009 and February to June 2010) on two cases of long-standing water conflicts between smallholder farmers and cities competing for water in the basin. The case studies are Shiri Njoro spring and the Nduruma river both located in the Kikuletwa catchment. The study involved interviews, and discussions with furrow (locally constructed irrigation canals) and river committees,[9] village leaders, farmers, city water authorities (technical manager), and staff of the Pangani Basin Water Office (PBWO). In Shiri Njoro spring we conducted group interviews with the chairmen and secretaries of three irrigation canals (total six members), and leaders of Shiri Njoro village irrigation canals association (total three members). We interviewed a representative of Moshi Urban Water Authority (the technical manager). The interviews focused on understanding the development of water use, the evolution of conflicts in the area and the strategies followed by the different users. Further downstream of Shiri Njoro spring, we interviewed five out of the seven irrigation committees (in total ten members, two from each furrow) and leaders of the overarching Kiladeda river committee (chairman and secretary). Shiri Njoro spring is one of the sources of the Kiladeda river.

In Nduruma, we conducted group interviews with village furrow committees of Bangata, Nduruma, Moivaro, and Midawe villages (total 28 members). We interviewed the representatives of commercial estates located in the mid-section of Nduruma sub-catchment: Old River Farm; former owner of Gomba Estate; manager of Dekker Bruins; the director and the irrigation manager of Arusha Blooms; and the

---

[9] River committees are water management structures created by the users to allocate and solve water conflict between users of a common river source (Komakech and Van der Zaag, 2011).

environmental and fertigation[10] officers of Kiliflora. We conducted a group interview with the leaders of Nduruma river committees (chairman and secretary) who were asked about their role in water allocation and management and how they relate with the Pangani Basin Water Board (PBWB).

To get a broader understanding of the administration of water rights we interviewed Meru and Hai district irrigation officers, the officer at the PBWO, the Nduruma ward executive officer and the Sokon II ward office chairman.

The study also benefited from unpublished sources. For the Shiri spring we reviewed letters, minutes of past meetings and reports compiled by the farmers, while for the Nduruma conflict secondary materials reviewed were mainly from Pamoja Trust (a local NGO based in Moshi) and PBWO archives. We also consulted relevant government documents, policies, water acts, and media reports on the water conflicts.

5.5   PANGANI WATER CONFLICT: CITY VERSUS SMALLHOLDER AGRICULTURE

In this section we present the water conflicts between the city of Arusha and Moshi and their rural neighbourhoods (Shiri Njoro and Nduruma). Currently water and sewerage services within the municipality of Arusha and Moshi are provided by fully autonomous public entities (Arusha Urban Water Supply Authority and Moshi Urban Water Supply Authority). Arusha city abstracts about 39,500 m^3/day, and Moshi about 24,000 m^3/day (EWURA 2010). However 26 and 32% of the abstracted water for Arusha and Moshi, respectively, is non-revenue water. To meet the demand of their growing population these cities increasingly appropriate water from sources already used by smallholder farmers. Before presenting the two case studies, we first provide a historical overview of the water rights administration in the Pangani river basin and in Tanzania as a whole (for more details, see Komakech et al. 2011b, Komakech et al. 2012a).

Historical context of water right development

The government's initial attempt to regulate water use by issuing water rights in the Pangani basin started during the colonial times. The colonial administration's intervention in water allocation in Tanzania as a whole was not about ensuring equity and sustainability of the water resources. Rather it was to support the interests of the commercial farmers and hydropower plants downstream. The Pangani and Rufiji basins were particularly designated for hydropower production and a special ordinance was prepared to protect such interest. In 1923, the British put in place the first Water (Utilization and Control) Act. Water users were required to acquire water rights, which were mainly issued to white commercial farmers who had settled along

---

[10] Fertigation is the application of chemical fertiliser and other products through an irrigation system.

the mid-reaches of Tanzania highlands. In the Pangani basin the commercial farmers settled on the slopes of Mount Meru and Mount Kilimanjaro forming what Spear (1997) called the 'iron ring' of land alienation. The Africans ("natives") were however allowed to develop a separate water allocation system building on local customs and traditions. The British created crown lands to be governed by statutory law and native reserves (land occupied by the Africans) to be governed by local law (Komakech et al. 2012a). This marked the beginning of a plural system of water governance in Tanzania's river basins. Local users have developed separate water-sharing arrangements at the level of an irrigation canal (between irrigated plots), between nearby irrigation canals along a river within one village, and between distant villages sharing a river (Komakech et al. 2011b).

The 1923 water act was subsequently amended by the colonial government in 1948 and 1959. The British declared absolute authority over water resources in the territory and introduced (nominal) water right application fees. In 1959, options of registration were extended to all water users including the Africans. National water officers were authorised to allocate and charge water right fees. These functions were delegated to regional offices. However, the British also put more emphasis on improving irrigation efficiency of farmer-initiated irrigation canals, which were believed to be wasteful.

The independent government of Tanzania later continued with the colonial policy of regulatory water allocation and management. All water resources were declared vested in the United Republic of Tanzania under the 1974 Water Utilization (Control and Regulation) Act. In later amendments the country was zoned into nine basins and Basin Water Boards were created to allocate and manage water resources. Under the influence of foreign donors, enforcing water rights became a mechanism for taxation. The World Bank particularly argued that rational water use could only be achieved through increasing economic water use fees. Low tariffs were stated to contribute to inefficient water use (World Bank 1996).

Irrigation improvement was recommended, since improved irrigation efficiency would release water from the agriculture sector to be used by highest-value uses, and in the case of Pangani and Rufiji basins, hydropower plants were located downstream. To support this point, the World Bank (1996) estimated the value of water in traditional irrigation at US$ cents 0.5 per m^3 of water and in improved irrigation schemes at US$ cents 3.0 per m^3 of water.

Following the recommendation of a rapid water resources assessment by the World Bank (1996), Tanzania embarked on a legal reform of the water sector with emphasis on regulatory water use. To regulate water use gates were constructed on irrigation canals abstracting water from major tributaries of the Pangani river. However, many gates were destroyed by farmers who did not agree with the state-led water right system. A revised National Water Policy was put in place in 2002 and passed into law by the Water Act of 2009. The policy embraces the principles of IWRM with the major goal of attaining equitable and sustainable management of the water resources. All water users are required to register and obtain permits indicating the purpose of use and the annual volume of water the users are entitled to. The permit holder must pay an annual water use fees calculated according to volume allocated and purpose of water use.

Current state-led water allocation and management in the Pangani basin are the responsibility of the PBWB. In accordance with the provision of the Water Utilization (Control and Regulation) Act of 1974 and its amendments of 1981, 1989, 1997, and 2009, an individual user, city authority or institution must apply for a water use permit. Officially, all rights applications are to be gazetted in a government newspaper for at least 40 days, during which all affected users have to be consulted and local district authorities must submit reports on the status of the water source. This includes recommendations from the District Agricultural and Livestock Development Officer, Regional Water Engineer, District Administrative Secretary, and District Executive Director. For large projects, clearance certificates of environmental impact assessment must be acquired from the National Environment Management Council. The PBWO also conducts studies on water availability. Based on the district department heads' recommendations and the water supply assessment report, the PBWB decides to grant or reject a water right application. In general, the application process can take months to years before a water right is granted (the PBWB only meets once every 3 months). Projects of national interest, e.g. water right applications for cities are, however, often expedited.

Moshi Urban Water Supply Authority vs. Shiri Njoro village farmers

In addition to three other sources (combined capacity 13,850 m^3/day), Moshi Urban Water Supply Authority (MUWSA) also obtains water from Shiri spring (10,150 m^3/day). Shiri spring is located about 7 km from Moshi town on the way to Arusha in the village of Shiri Njoro, Hai district (Figure 5.3).

The spring feeds the Kiladeda river and forms parts of the river network originating from the slope of Mt. Kilimanjaro flowing into the Pangani river. MUWSA is the biggest water user but farmers from Shiri Njoro village rely on the spring for small-scale irrigation, and domestic and livestock needs. Farmers have constructed three irrigation canals (Kitifu Mashariki, Kitifu Kati, and Kitifu Magharibi) which they use to irrigate yams, bananas, maize, coffee and vegetables.

PBWB has so far issued six volumetric water use rights on Shiri spring (collective and individual): MUWSA 116 l/s; chairman Kitifu Mashariki 30 l/s; chairman Kitifu Kati 30 l/s; chairman Kitifu Magharibi 30 l/s; Elisa G Mallya 1 l/s; and J.P. Muro 1 l/s. The total allocated abstraction of the spring flow is thus 208 l/s, and the PBWB estimated the average yield of the spring at 218 l/s, so about 10 l/s is left to flow into the Kiladeda river.

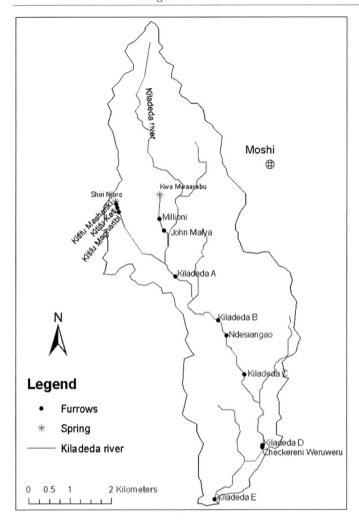

Figure 5.3: Kiladeda river sub-catchment, springs, furrow intakes and location of Moshi town.

Historical evolution of Moshi city control of Shiri spring

In the 1950s, farmers established the three canals drawing from the Shiri spring for supplemental irrigation during rainy seasons and full irrigation in the dry seasons. To construct the canals, farmers sought permission from the area chief. About the same time the British colonial administration also constructed a water supply line (10 inch pipe abstracting about 56 l/s) for Moshi town on the same source. Moshi town's intake was located upstream of the existing Shiri Njoro village canals. According to the farmers there was no water conflict but once in a while they would experience water shortages.

Box 5.1: Evolution of Shiri spring water conflict (Source: based on Shiri Njoro farmers file).

1994	Kilimanjaro Regional Water Engineer informs Shiri Njoro village about the new water project of MUWSA. The villagers responded that additional abstraction by MUWSA will aggravate water shortages in the village.
1998	Farmers create the Shiri Njoro canal association and write to MUWSA that the new project will affect the 411 households dependent on the water for domestic use and irrigation. The farmers apply for a formal water right for the entire spring water (300 l/s) but the PBWB allocates the three canals 30 l/s each.
1999	The village canal association petitions the Director of Development of Hai district about the additional allocation to MUWSA. Lyamungo Division Secretary writes to MUWSA advising the authority to dialogue with the village.
2000	The Regional Water Engineer holds several meetings with the village and MUWSA. The PBWO clarifies that MUWSA will take 116 l/s and that with a spring yield of 300 l/s there will be sufficient water for the canals. PBWO directs MUWSA to provide domestic water through standpipes to the village.
	When MUWSA fails to abide by the agreement, the farmers petition the Kilimanjaro Regional Commissioner, who calls for proper research on water availability. PBWO finds that the spring yield was 218 l/s, total abstraction 209 l/s, leaving 9 l/s as inflow to Kiladeda.
2001	The village Executive Officer of Shiri Njoro complains to the regional water engineer of Kilimanjaro that MUWSA now abstracts all the water from the spring and uses two pipelines. He states that the villages are preparing to destroy the MUWSA pipelines.
	The Regional Water Engineer states that field measurement carried out in March (start of the rainy season) found that the available water at the spring had reduced from 218 to 181 l/s. The District Commissioner tells MUWSA to remove the old pipeline.
	MUWSA refuses to ration water and continues to abstract more than allocated, arguing that since domestic water takes priority, it is up to the village farmers to reduce their use.
2002	The District Commissioner contacts the Kilimanjaro Regional Commissioner, stating that the main problem was that the new MUWSA pipeline takes 127 l/s and that only the first canal (Kitifu Magharibi) receives water. Downstream farmers react by destroying the intake of the first canal. The PBWO intervenes and tells the farmers to rebuild the canal intake.
2003	MUWSA continues to abstract more water than allocated. The village farmers refuse to pay the annual water fees.
2004	The village canal association chairman writes to the PBWO, complaining that the three village canals have failed to get their allocated 30 l/s per canal and that cash crops have dried up.
2007	The Shiri Village Executive Officer writes to the PBWO complaining of over-abstraction by two individual farmers, Elisa Mallya and Lt. Col. Muro. The farmers invade their homes, destroying water infrastructures.
	The PBWO conducts flow measurements and finds that MUWSA abstracts 120 l/s;

> the first canal (Kitifu Magharibi) 17 l/s; the second canal (Kitifu Kati) 13 l/s and the third (Kitifu Mashariki) 15 l/s. Mallya and Muro abstract no water because their intakes are destroyed. PBWO warns that damaging the infrastructure of other users is against the Water Act of 1974 section 33 (2).

In 1994, MUWSA applied for an additional allocation of 68 l/s from Shiri spring and was granted 60 l/s by the PBWB. The new allocation thus increased total water right of MUSA on Shiri spring to 116 l/s. The Regional Water Engineer of Kilimanjaro informed the Shiri Njoro village chairman about the additional allocation. The chairman however responded by highlighting water shortage in the village and stated that during a village meeting farmers objected to the additional water allocation to MUWSA. Several communications, meetings and confrontations have since taken place and still continued when fieldwork was conducted. No solution has been found according to the chairman of the village canal association. The village canal association kept records of all meetings, letters, water conflict events and reports related to the spring water conflict. We were provided access to these records by the secretary of Shiri Njoro village canal association. Shiri farmers are now looking for a lawyer to argue the case in court. Box 5.1 presents a review of the evolution of the Shiri spring water conflict.

Summary of Shiri spring water conflict

Shiri spring is a small water source but the issuing of water rights has not led to orderly use or even increased efficiency. The Shiri spring case highlights the challenges of administering a formal water right system in the Pangani river basin. The smallholder farmers are willing to jointly manage and allocate the waters of Shiri Njoro. MUWSA has been uncooperative and PBWO has been unable to resolve the emerging conflict. The spring flow is over-allocated. Fixed volumetric water rights were issued based on the assumption of a constant spring yield of 218 l/s. However, recent field measurements by PBWO indicate that the spring yield is frequently much lower. MUWSA continues to abstract over and above its allocated share including during periods of low flow, leaving the villages with about half their formally allocated right. MUWSA technical manager states that since the government gives first priority to domestic water use, the city does not feel obliged to reduce its share when the spring yield decreases. When asked if they could compensate the farmers for the lost income, the manager said that the city already does that by paying the annual water user fee to the PBWB. The case also shows that the water scarcity created by Moshi city causes internal water struggles between irrigators.

The following section presents a similar water conflict in the Nduruma sub-catchment. Unlike Shiri spring, Nduruma water is used by both smallholder and large-scale commercial farmers.

Arusha Urban Water Supply Authority vs. Nduruma water users

Nduruma river crosses eight administrative wards of Arumeru district, with its headwaters located within a protected forest reserve on the slopes of Mount Meru (Figure 5.4).

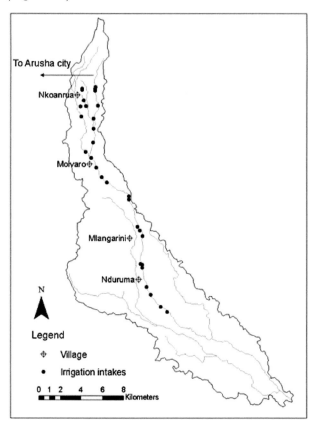

Figure 5.4: Nduruma sub-catchment, irrigation intakes and villages.

The highlands are occupied by smallholder farmers who maintain irrigation canals and grow crops like beans, coffee, bananas and potatoes. The midlands are the most intensively farmed along the Nduruma river. Here the majority of the farmers have large commercial estates (mostly foreign-owned), first created by the colonial government (Spear 1997) and later privatised in the 1990s. Crops grown include coffee, flowers, horticultural crops, fruits, bananas, maize and beans. The lowlands were recently settled by people escaping land shortage in the highlands and former estate workers. The first group of immigrants moved into the area during the colonial period and a second group arrived in the 1970s mainly stimulated by the national government's village resettlement programme. The inhabitants came from different groups and most are smallholder farmers. Crops grown include maize, beans, banana, cassava, pigeon peas and horticultural crops. Also a significant number of freely grazing livestock are kept. This zone experiences extreme water shortages during the dry seasons.

Irrigation along the Nduruma river has been practised for over 200 years (Spear 1997). Nevertheless, agricultural intensification started during the colonial period when commercial farmers (Germans, Greeks and British) settled in the area. At the end of colonial rule, the majority of the irrigation canals in the Nduruma highlands belonged to Africans, the midlands to Europeans and the lowlands to Africans (Komakech et al. 2012a). The situation is nearly the same today. The Nduruma river is over-committed to agriculture such that it now only flows for part of the year (Figure 5.5). The estates were first issued water rights during colonial times. These were later reviewed by PBWO in the 1990s.

However, with the increasing population of Arusha city and the booming tourism and mining industries the water demand of the city is on the rise (Komakech and Van der Zaag 2011). The city therefore was forced to look for water from the surrounding rivers, including Nduruma. Here, Arusha city built an intake located upstream of all the existing water users (Figure 5.5). For farmers, the arrival of the Arusha city water authority could only exacerbate the competition over scarce water.

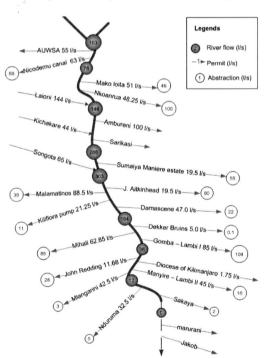

Figure 5.5: Flow diagram of Nduruma river, showing water abstractions and inflows as measured by PBWO in November 2003 (dry season). Groundwater inflow was not accurately determined, hence the figures do not add up (Source: Komakech et al. 2012a).

Historical evolution of Arusha city water control in Nduruma catchment

The struggles over land and water access in the Nduruma sub-catchment started during colonial times when local farmers were dispossessed of their lands and water resources (Spear 1997, Komakech et al. 2012a). The commercial estates were issued water use permits while Africans were allowed to use water according to their customary arrangements. In 1968, the commercial estates formed the Nduruma Water User Association to strengthen their negotiation position. However, around the same time the government started its resettlement programme (the villagisation programme). Most of the coffee estates collapsed and have only recently been revitalised. Since its creation the Nduruma water user association has never really functioned and is not known to the PBWB or to the district authorities. However, a river committee has been created by the farmers to oversee the water allocation between the midland and lowland farmers (Komakech et al. 2012a). This river committee attempts to reduce water use by the estates from 24 hours, as stipulated in the state-issued water right, to 6-9 hours per day. The weakness of the Nduruma river committee is that its membership excludes highland users and only encompasses commercial estates and the downstream small-scale irrigators. The committee members are in effect distributing amongst themselves the water that the highland villages were unable to use (Komakech et al. 2012a). The river committee leaders explained that they lack the power to reach out to the representatives of the upstream users. This is because the upstream villages have water allocation arrangements that lack formal structures for downstream villages to engage with. The highland irrigation canals have a committee which is responsible only for maintenance and allocation of water to individual farmers (Komakech et al. 2012a). Representatives from two midland estates (Dekker Bruins and Tanzania Flowers) confirmed that on their own it is not possible to discuss water-related issues with upstream users; they need the district office to act as an intermediary. They also explained that even when the district office intervenes their influence only lasts about a week, after which upstream users stop cooperating.

In 2001, the PBWB granted AUWSA a 55 l/s water use right on the Nduruma river to supplement the city's growing domestic water needs. AUWSA gets most of its water supply from springs, boreholes and river sources located within the Themi sub-catchment (Komakech and Van der Zaag 2011). The Nduruma permit gave priority of allocation to Arusha city which is located 40 km outside the Nduruma sub-catchment. It also allowed the city to construct its pipe intake upstream of existing users, creating a locational advantage that coincides with the priority status of the permit. Existing users were not involved or consulted in the issuance of this water right as stipulated in the water act. In 2003, AUWSA started constructing the Nduruma water pipeline, but conflicts between AUWSA and the various downstream users soon erupted. This included a violent riot in October of that year which temporarily put the AUWSA project on hold.

Box 5.2: Evolution of AUWSA versus Nduruma farmers' water conflict.

2001	PBWB issues a water permit to AUWSA to abstract 55 l/s of water from the Nduruma river. The local farmers and estates, having state-issued water rights, are not informed.
2003	AUWSA starts constructing a pipeline at the source of the Nduruma river. In October 2003, farmers attack the contractor employed by AUWSA, his car and 300 culverts are destroyed. Construction is stalled.
	Commercial farmers with water rights issued by PBWO get involved. The Gomba estate managing director sends a letter of objection to PBWO, highlighting the importance of Nduruma to Tanzania's largest flower, fresh vegetables, and horticultural farms. He argues that water availability in Nduruma is at a point where an upstream user cannot access his or her full water right without affecting the allocated rights of other users downstream. The letter is copied to the President of Tanzania and the Ministers of Finance, Agriculture and Food Security, and Water and Livestock Development.
	The chairman of the reconstituted Nduruma Water User Association appeals to the Minister of Water and Livestock Development, stating that contrary to the Water Utilization Act of 1974, they were not informed of the AUWSA water project.
	The Principal Water Officer, Ministry of Water and Livestock Development, urges PBWO to conduct a water assessment and to encourage dialogue between the users. The Basin Water Officer meanwhile responds to the Gomba estate's director that it will review all existing water rights and make variations to all right holders as the available water is inadequate to satisfy all rights.
	PBWO conducts a (new) water assessment and through an extraordinary PBWB meeting reduces all water permits by 20% and all permits are declared provisional rather than permanent. AUWSA is allocated 44 l/s but permitted to increase its use to 55 l/s during the rainy season.
2004	IUCN and Pamoja (a local NGO), through a 'dialogue on water project', try to get involved but are unsuccessful in mediating the water conflict.
2006	Gomba estate closes down partly due to lack of water security. Its property is sold to Arusha Municipal council, who intends to establish a satellite township. AUWSA completes its water project. The site is guarded by a private security company and locals are denied entry to the forest without security escorts.
2009	Young farmers from Bangata and Nkoanrua villages attack the AUWSA abstraction point with machetes.

Against the threat of this new pipeline, all midland and lowland users – villagers and estates alike – found themselves momentarily on the same side. The Gomba Estate manager (Mr. Michael Chamber) knew that Arusha city would reduce his water supply in the dry season, when he needed it most, far below the flow levels stated in his water right. He petitioned the district commissioner on the new allocation to Arusha city and when the district commissioner refused to respond to his complaints, Chamber decided to engage with the villagers. According to Chamber, he discovered that there was an inactive Water Users Association (WUA) that had been founded in 1968 but not registered with PBWO. He thought that if this extinct WUA were legally recognised it could add potency to his arguments against the pipeline construction. He helped revive the Nduruma WUA with Mr. William Nassari of the downstream Nduruma Village acting as chairman, and Michael Chamber serving as secretary. The District Irrigation Office interprets these events as Gomba Estate using the downstream smallholder farmers to protest against AUWSA. He claims that Chamber even hired trucks to transport the angry villagers up Mount Meru to the source of the Nduruma, were they began to riot. A publicity campaign was later launched by AUWSA in an attempt to sensitise the downstream users about the importance of Arusha's domestic water project. Security was increased at the water abstraction site, including the construction of a permanent police station to monitor the area. PBWO and the district authority organised meetings in each ward and the villagers were strongly dissuaded from continuing to participate in "Chamber's WUA". The reconstituted WUA meetings have since stopped and there has been no effort to revive it since 2004 when the AUWSA pipeline became operational, and downstream users try to cope with increased water shortages, especially during the dry seasons.

Box 5.2 presents the historical evolution of the conflict between AUWSA and Nduruma water users.

More recently, in 2011, PBWB and its development partners (International Union for Conservation (IUCN), Pamoja Trust and Stichting Nederlandse Vrijwilligers (SNV Netherlands Development Organisation) created a separate Nduruma river committee that is supposed to link all the water users in the sub-catchment (Komakech and Van der Zaag 2013). However, the new Nduruma committee has not been operational as yet, as the upstream farmers remain unwilling to cooperate with the downstream users.

Summary of Nduruma river water conflict

Like Shiri Njoro farmers, existing water users in Nduruma were not consulted or compensated. To this day, farmers remain dissatisfied with the situation. They claim that AUWSA abstracts far more water than their nominal allocation and that AUWSA does not respond with any sympathy to the farmers' complaints of scarcity in the dry season. Their position is also shared by the Arumeru district executive director. According to the director it is unfair for AUWSA to tap Nduruma water without making any contribution to the villagers using the same water source. The director argues that AUWSA earns a hefty income from Nduruma water so it would be fair to pay royalties to villagers for the management and protection of the water resources. The AUWSA managing director, in contrast, claims that since water resources in Tanzania are government property it is the responsibility of the government to decide how best it is used. Occasionally the frustration of the villagers is manifested in violent ways. In January 2009, a band of young farmers from Bangata and Nkoanrua villages attacked the AUWSA abstraction point with machetes. Interestingly, in 2003 the PBWO revised all water rights downwards by 20%, and further made the water right of the city flexible: AUWSA was allowed to increase its use to the originally allocated 55 l/s during the rainy seasons only.

Farmers interviewed from the villages of Mako Loita, Bangata, Midawe, Mlangarini, and Nduruma all stated that they would like PBWO to institute a system that obliges AUWSA to reduce its allocation during the dry season.

The people affected most by AUWSA water use are farmers using the nearest three irrigation canals, i.e. Nicodemu, Mako Loita and Nkoanrua. Downstream of these irrigation canals there are more springs and streams joining the river. Midland and lowland farmers have initiated a rotational allocation system (domestic and livestock uses inclusive) that is negotiated on a seasonal basis (Komakech et al. 2012a). Although commercial estates and smallholder farmers tried to create a front by reconstituting the 1968 Nduruma Water User Association, they were not able to stop AUWSA from taking control of Nduruma water. AUWSA was the last user to arrive in Nduruma but has now the first call on the water. Water use by Arusha has had negative local socio-economic consequences: one estate (Gomba) closed down and many villagers lost temporary or permanent employment opportunities.

5.6 DISCUSSION

Water distribution in Pangani is characterised by local investments in water infrastructure and local distribution rules that evolved over time. These infrastructures and customary allocation rules take water variability into consideration. In many places, irrigation areas expand and contract in sync with water availability and the allocation rules also change with the seasons (dry and wet seasons). In the nearby Themi river, for example, water is reserved for domestic and livestock use during certain periods of the dry season (Komakech and Van der Zaag 2011), while in Makanya catchment downstream farmers are encouraged to borrow land in the upstream parts of the irrigation command areas during dry seasons (Komakech et al. 2012d). Lankford and Beale (2007) found a similar arrangement in the Usangu basin, also in Tanzania, and report that 20% of the maximum area could always be irrigated and that the maximum area can only be served during exceptionally wet years.

Both case histories presented in this chapter show the existence of a hybrid, plural legal situation: prior customary uses co-exist, and sometimes clash, with formal government laws first initiated by the colonial administration and later revived by the independent government of Tanzania. Officially, the Pangani Basin Water Board and Office is responsible for the allocation of water use rights and management of water resources of the basin.

The PBWO relies on reports and assessment studies to allocate water rights, but does not monitor actual water use. The board's staff only moves around to collect the annual water user fees. In many places, water abstraction exceeds the allocated amount and the increased use of mobile water pumps by dispersed smallholder farmers particularly renders the water administration system ineffective. We observed a complex water-sharing arrangement, especially in coffee estates on the slopes of

Mount Kilimanjaro. These estates were originally managed by rural cooperative societies, which have now leased their estates to private investors with the mandate to grow coffee. However, these investors are also subletting the farms to other private investors who mostly grow high-value crops: flowers, tomatoes and green beans for international markets that incidentally also require to be irrigated. The complex arrangements mean that those private investors without water rights collaborate with those with water rights to abstract more water. PBWO does not have the means of verifying the water use by the estates. An interviewed officer of PBWO said that "if there is no problem, you do not need to disturb the equilibrium or the flow of the system. It is difficult for us as PBWO to check all water use; we only get involved when there is conflict."

Cities acquire their water rights from PBWO. The smallholder farmers use their prior customary uses and governance arrangements to claim access right, while cities use formal law (particularly priority allocation to domestic uses) to gain control of water. The smallholder farmers are officially obliged to apply for formal water rights as well, but in practice only some irrigators with individual canals and pumps have done so, but none of the furrow irrigators. This may possibly be explained by the fact that the latter are not as administration-savvy compared to cities and commercial farmers or that they find the statutory system lacks legitimacy. Although smallholder farmers were the first to start using the water sources, they are increasingly being made the last by PBWO which gives allocation priority to cities.

Water transfers to cities in the Pangani river basin may be categorised as permanent and outright (cf. Molle and Berkoff 2009). The mechanism used to reallocate water to the city of Arusha and Moshi followed a combination of administrative decision and stealth. The formal water law requires that before issuing a water use right, all potentially affected parties be consulted and existing or potential water conflicts should be resolved before any new allocation can be made (Tanzania 2009). In both Shiri spring and Nduruma, farmers were not properly consulted by the basin water board/office. Farmers using Shiri spring and Nduruma both contested the appropriation (e.g. by rioting, by involving various department heads and political leaders), but they were not able to stop the powerful city water authorities from gaining control of the water.

The two case studies highlight aspects not often mentioned in the literature on inter-sectoral water transfers. First, unlike other cases (see Loeve et al. 2004, Bhattarai et al. 2005, Hearne 2007, Celio et al. 2010), Pangani is a basin where 80% of the users are smallholder farmers who have invested significant time and labour in the construction and maintenance of irrigation canals. In such a setting reallocation to cities does not only deprive farmers of water but may also render their long-term hydraulic property investment (partly) obsolete. Smallholder farmers rely on the irrigation canals to mitigate the impact of agricultural drought, and realise their food security. In addition, these canals serve other purposes as well, including livestock watering, construction (brick-making) and, importantly, domestic uses. This domestic use is often overlooked, yet should also be accorded priority. Further, it could be noted that furrow water use is fully consistent with the government policy that aims at eliminating rural poverty.

Second, the water policy and act give priority to registered domestic water uses and cities and only state that the other uses will be allocated taking into consideration the economic and social values. In the Pangani, cities are given first priority and the other users get a proportional reduction in the allocation (illustrated in the Nduruma case where the basin board reduced all existing allocation by 20%). There are no planned measures for unregistered uses and registration does not change the priority allocation of the furrow. Although the new water act recognises customary water users and obliges them to formally register their use, it does not have a mechanism for compensation for lost livelihood in case the water is reallocated to new users (Tanzania 2009). This in a way explains why the smallholder farmers do not bother to register. In our view, the recognition of customary uses should be grounds for compensation in case existing uses are impaired. This would create incentives for any newcomers to look for alternatives before appropriating water from existing users.

Third, the current water right licensing system does not match the basin reality. It is a system that allocates fixed water use rights on the assumption that an average level of supply exists. However, water supply in the Tanzania river basins is highly variable due to unpredictability of rainfall and recurrence of droughts and floods (see Lankford and Beale 2007). In such a situation, and in the absence of water storage infrastructure, it is the low flows during the dry seasons that pose allocation challenges. The water act makes a provision for the revision of water use permits in any specified area where the available water is insufficient to satisfy all permits. But the process does not work fast enough especially for dry-season scarcity. Formal water rights could benefit from local water allocation systems (cf. Horst 1998). In Pangani, local farmers have developed flexible water allocation rules, schedules or abstraction turns that are renegotiated on a seasonal basis within the area served by irrigation canals and also among intakes along the river (Komakech et al. 2012d). Lankford and Mwaruvanda (2007) propose one such framework: a legal-infrastructure framework that integrates formal water rights and customary agreements by establishing a wet-season volumetric cap and a dry-season proportional cap for all allocations (Lankford and Mwaruvanda 2007). The decision to award AUWSA a water right that may increase during the wet season by 20% may be seen as a first step in this direction. Negotiated water allocation is driven not only by economic power but also by other values, including social values and interests, which are allowed to be heard in the negotiation process (see Molle 2004 for a discussion on negotiated water allocation). Given the context of water use development and variability in the basin, negotiated water allocation can potentially mitigate water conflicts and reduce potential downstream impacts.

Finally, given the power inequality between the city authorities and smallholder farmers, the capacity of the PBWO to set, monitor and enforce fair water allocation rules is very important. In the Pangani basin, the capacity (technical and financial) to enforce and monitor compliance with water allocation rules is still weak. City water authorities exploit this gap and abstract more than they are allocated but go unpunished. In both Nduruma and Shiri spring, the rural-to-city water transfers have had significant downstream impacts. Shiri farmers claim to have lost entire coffee plantations and fish ponds and that conflicts over water among farmers has increased. Three irrigation canals in Nduruma - Nicodemu, Mako Loita and Nkoanrua - despite being among the oldest users do not receive sufficient water most of the time. Gomba estate, that was very vocal and outright in objecting against the allocation to Arusha city, collapsed. The impact of this water appropriation, however, disproportionately affects the smallholder farmers. While the large commercial farmers can leave with some of their investments and look for alternatives, the smallholder farmers lose all their investments with much fewer alternatives and are forced to rely on marginal rain-fed agriculture, or have to join the peri-urban poor.

5.7 Conclusions

This chapter described and analysed appropriation of water from smallholder irrigators by cities in a river basin that is becoming water-stressed. In such a stressed river basin there is a need for the state-based water rights allocation system to have legitimacy. However, in practice the rights system as administered by many governments may in fact provide the legal means for some actors to dispossess existing users. Powerful cities discussed in this chapter selectively use the law to gain leverage over water control.

In general, water appropriation and transfer to growing cities in the Pangani basin and other basins in sub-Saharan Africa is an ongoing process (Gupta and Van der Zaag 2008). These transfers take place in a context where prior investments (infrastructure and institution) have been made by smallholder farmers. This is in contrast to situations where the state has invested heavily in hydraulic infrastructures (e.g. in storage, conveyance) and where there is a strong institutional capacity and where prior uses are recognised.

Water reallocation to the cities of Arusha and Moshi was achieved through a combination of administrative decision and stealth. The justification was that domestic water use has the highest priority as stipulated in the water policy and law. However, in issuing priority rights to these cities, the domestic needs of the smallholders were not considered, let alone the fact that the farmers rely on the irrigation canals to realise their food security and livelihoods.

Water capture of rapidly expanding cities seems inevitable. It is therefore essential that suitable options for water allocation be applied to minimise the potential negative impacts of the agriculture-to-city water transfers or water allocation in general. We propose the following for Pangani, but this may also be applicable to other basins in semiarid areas particularly in sub-Saharan Africa. First, the state-administered water rights system could incorporate a proportional allocation system that comes into operation during periods of water scarcity (see Van der Zaag and

Röling 1996, Horst 1998). By proportional allocation we mean a situation whereby the state-issued water right entitlements are reduced in proportion to the expected/observed shortfall in river flow. This could be negotiated by the users at the river level - for instance in Nduruma the river committee currently negotiates rotational allocation between smallholder farmers and estates, the latter having state-based water rights. Presently, the Tanzania water right system is based on average flows, without recognising the normal flow variability (in the absence of storage reservoirs which could buffer such fluctuations). Formal water rights could benefit from the time-tested local water allocation system that does take variability into account (Komakech and Van der Zaag 2011), and from the suggestions made by Lankford and Mwaruvanda (2007) and Lankford and Beale (2007) on a legal-infrastructure framework.

Second, the PBWO capacity (technical and financial) to enforce and monitor compliance with the state water allocation rules is weak. Currently the basin water board does not monitor water allocation; it only gets involved in cases of conflict. This may explain why the water authorities in both cities continue to over-abstract water, leaving little to smallholder farmers. The institutional capacity to monitor and enforce agreements has thus to be strengthened. This could be done by recognising local arrangements (e.g. river committees active in Nduruma and Kiladeda rivers) and make them responsible for the negotiation of water allocation (Komakech and Van der Zaag 2011). It also means that any water transfers or allocations to cities would be negotiated by these river committees. The PBWO would then focus on backstopping with technical information.

Finally, when there is good knowledge of the water resource, it may be possible to introduce a system of water trade or payment for lost benefits in the form of lease or option contracts that would only come into operation in times of scarcity (see Howitt 1998, Characklis et al. 2006). In the Pangani, smallholder farmers are not being compensated for when their water is appropriated by cities. Since the water law does recognise customary uses, allows water trade and also allows the basin water board to attach any condition to a use permit (including compensation to any other person), it is theoretically possible to institutionalise a compensation scheme that recognises the prior water use and farmers' investments in water infrastructure. This could be grounds for compensation of smallholder farmers and may create incentives for newcomers to look for alternative water sources, invest in additional storage options or invest in demand management and leakage control. In so doing the first would not have to come last.

PART 3: REDISCOVERING LOCAL WATER

GOVERNANCE IN THE PANGANI BASIN

To any farmer, owning 1000ha of land without irrigation water is the same as having a big bowl of dust; 1ha of land with sufficient irrigation water to grow crop is better than 1000ha without water. Having access to sufficient water to grow crop in 1ha of land is far much better than farming 1ha without water security; you will harvest. However, if the farmer applies fertilizer and pesticides to his/her 1ha of land with sufficient irrigation water he/she will harvest much more crop (Personal communication: Xanfon Bitala, 2013, Tanzania).

It is apparent from the discussions in Part 2 that local efforts are not being fully integrated into government water sector institutional reforms, leading to mismatches between the newly created organizations' and locally evolved ones. Understanding why local arrangements emerge, and function, even in situations of water asymmetry and inequality is therefore of importance. In this section, I address the first research sub-objective: to understand the mechanisms that drive cooperation at the local level (e.g. turn taking in villages – between farmers sharing a furrow, between two furrows, between neighbouring villages, between distant villages and within a catchment). I use a combination of concepts to describe and analyse the process of local institutional change and its interface with state-led reforms.

In Chapter 6, I discuss a situation where water asymmetry, inequality and heterogeneity dynamically interact and give rise to interdependencies between water users which facilitate coordination and collective action. There is also a positive linkage between inequality in access to land to the emergence of collective action e.g. people in access to more land are more likely to assumme the leadership role needed for continuous maintenance of the irrigational canal. This finding however is limited to a small spatial scale involving one village. I posit that at a very local spatial scale there may be inhibitions of unilateral action due to the social and peer pressure. This is because at a small spatial scales e.g. of a village, it is more likely that enough people (critical mass) are assured that others will cooperate with respect to a resource use, which motivates the emergence of collective action. It is therefore likely that spatial and social proximity may be a necessary condition for collective action in water asymmetrical situation to emerge.

At a larger spatial scale (Chapter 7), river committees have emerged to solve coordination challenges over water use between villages, and between large- and small-scale farmers. A river committee brings together small and large scale farmers, districts, villages and wards thereby bridging hydrological and administrative boundaries. Although a clear boundary definition is identified in the literature as one of the key conditions for the survival of local self-governing institutions, I find that the boundary definition adopted by the users at the river scale is dependent on how far the downstream users claim to water can go upstream. The users employ an ambiguous boundary definition to allow for future negotiations. Location (water) asymmetry as observed at the level of irrigation canal (Chapter 6) and not heterogeneity makes upstream users less willing to share water. I observed that distant upstream villages tend to use their location advantage to claim ownership which they consider synonymous with a license to use more water. Thus at larger spatial scales the anonymity between users makes it more difficult to initiate and maintain collective action.

At the interface between state-led and local arrangements, I observe situations of struggle over control of dwindling supply. The state's attempt to formalize water allocation systems raises the question which institutional arrangements are considered legitimate. In Chapter 8, I present a case of a struggle over water rights between large- and small-scale farmers using one river. Legitimacy is at the centre of this water struggle. Large estates that are in the possession of state-issued water rights, must adapt to local concepts of water sharing in order to secure access to irrigation water. State water rights do not translate into automatic access to water at the local level. Adhering to local norms does legitimize one's claim to water. Although the large estates can on paper claim water access using state-issued permits they adopt a variety of strategies - engage with the local system and negotiate a fair rotational allocation system; or switch to less visible groundwater. One of the strategies used by large-scale farmers is to build social reputation with the small-scale farmers thus increasing the chance of cooperative behaviour. These larger water users thus play an important role in sustaining collective action. Their behaviour may be explained in that they may have a larger stake in the water and thus in collective action. In addition, these larger estates are dependent on the farming households for labour, and the latter for wages. Their mutual dependence may explain why a negotiated status quo between small-scale and large-scale water users in a situation of severe water scarcity, can endure.

Chapter 6

THE DYNAMICS BETWEEN WATER ASYMMETRY,

INEQUALITY AND HETEROGENEITY SUSTAINING

CANAL INSTITUTIONS IN THE MAKANYA

CATCHMENT, TANZANIA[11]

6.1 ABSTRACT

It has been suggested that the collective action needed for integrated water management at larger spatial scales could be more effective and sustainable if it were built, bottom-up, on the nested arrangements by which local communities have managed their water resources at homestead, plot, village and sub-catchment levels. The up-scaling of such arrangements requires an understanding of why they emerge, how they function and how they are sustained. This chapter presents a case study of local level water institutions in Bangalala village in the Makanya catchment, Tanzania. Unlike most research on collective action in which water asymmetry, inequality and heterogeneity are seen as risks to collective action, this study looked at

---

[11] Based on Komakech et al. 2012. *Water Policy* 14(5), 800-820

how they dynamically interact and give rise to interdependencies between water users which facilitate coordination and collective action. The findings are confined to relatively small spatial and social scales, involving irrigators from one village. In such situations there may be inhibitions to unilateral action due to social and peer pressure. Spatial or social proximity may thus be a necessary condition for collective action in water asymmetrical situations to emerge. This points to the need for further research, namely to describe and analyse the dynamics engendered by water asymmetry, inequality and heterogeneity at larger spatial scales.

6.2   INTRODUCTION

Canal organizations in the Makanya catchment, Tanzania, have been managing the allocation of water for multiple uses between households for more than fifty years. Water is used for domestic purposes, construction (brick making), the watering of animals and for irrigation. The latter use will be the focus for this chapter as it is the largest water consumer by far, and also because (in other canal systems reported in the literature) farmers have been able to negotiate water-sharing arrangements at plot, village and sub-catchment levels (see Gray 1963, Fleuret 1985, Wade 1988, Ostrom 1990, Grove 1993, Ostrom 1993, Adams et al. 1994, Adams et al. 1997), overcoming collective action problems related to water provisioning and allocation in many parts of a catchment. It has been suggested that understanding why these arrangements endure even in the face of access inequality could form the basis for integrated water management at larger spatial scales (Van der Zaag 2007).

In this chapter we propose that local water institutions can endure over time because three phenomena, namely water asymmetry, inequality and heterogeneity, dynamically interact and create interdependencies between water users that may foster coordination and collective action concerning a particular water resource. We provide theoretical arguments for this proposition and illustrate it with a case study. In the concluding section we show the relevance of this proposition for contemporary water resources management and formulate a research agenda. In this introduction we briefly summarize our proposition, focusing first on inequality, then on water asymmetry, and on heterogeneity.

The use by many appropriators of a common pool resource such as a fishery or a groundwater body (aquifer) requires coordination. It is widely accepted that coordinating many dispersed users poses a challenge and, in many instances, proves ineffective or impossible. It has been shown that inequality among appropriators, for example manifested by unequal amounts of the natural resource used, may sometimes lead to collective action. This is because the large resource users tend to be willing to contribute more to collective action, which may in specific cases be sufficient to allow some free-riding of small-scale users (see Olson 1965, Baland and Platteau 1999).

Water flowing through a river or a canal can also be understood as a common pool resource. However, in such a case there is an additional complication, namely that surface water normally flows in one direction only. This unidirectional effect creates an asymmetrical situation: the action of upstream appropriators to use or refrain to use water influences the ability of downstream users to do the same, but not vice versa. This water asymmetry adds a new dimension to the coordination problem, as upstream users are not interested in collaborating.

Here, collective action and coordination is less self-evident than in more symmetrical common pool resources such as fisheries or aquifers. Situations where, in the absence of a regulatory authority, such coordination nevertheless occurs thus merit careful analysis. We hypothesise that inequality will not be a sufficient factor to induce coordination in the case of flowing water, but that the interdependence of the resource users may be an additional requirement, and that interdependence is likely to arise out of the differences, other than inequality, between the resource users. This diversity we denote as heterogeneity.

Heterogeneity is thus defined here as the diversity in character or content of a particular entity among a group of users. Heterogeneity can arise from differences in soil types, differences in (micro) climate, differences in crops grown, irrigators having plots in different parts of the command area, socio-cultural differences and kinship relationships between irrigators in different locations of the command area. Such heterogeneities may create interdependencies among users because individual users specialise, e.g. in producing certain crops, which may be bartered with other irrigators producing different crops, or because irrigators may have multiple loyalties. Interdependence is defined as a situation where two or more people depend on each other in more than one way; it highlights the hydrological, social and economic feedbacks of actions or inactions of users in a canal or catchment. Interdependence is here considered to be an emergent property of heterogeneity.

Thus, while water asymmetry may impede coordination or cooperation over the allocation of water resources, this might be counterbalanced by: (i) the interdependencies arising out of heterogeneity within the user group; and (ii) inequality in terms of land or water use, which may allow large users to invest more than proportionally in coordination efforts. In such a water system where water flows from upstream to downstream, this dynamic interaction may influence and reconfigure power relations. The status quo can be interpreted as being the outcome of a delicate balance that is constantly being challenged and has to be re-enacted continuously; and in this process water institutions evolve and endure.

A typical example where inequality, asymmetry and heterogeneity coalesce is the case of canal maintenance in many farmer-initiated irrigation canal systems, as reported by for example by Martin & Yoder (1988), Boelens & Davila (1998), and Manzungu *et al.* (1999). Interdependence between water users may arise when labour requirements for canal repairs is high (Martin and Yoder 1988, Boelens and Davila 1998, Manzungu et al. 1999). This often occurs in the head end of an open canal system (at a river diversion, intake works, or the first stretches of canal that may cut through rough terrain before reaching the command area), and in such cases the irrigators in the head end, though in an advantageous hydraulic position, depend on the cooperation of their tail-end counterparts. Seasonal work parties are sometimes ritualised and may reconfirm the membership of the common canal and the claim to water for all. This composite of factors appears to sustain such systems despite of a clear water asymmetry and marked differences between irrigators.

The remainder of this chapter is organized as follows. Section 6.3 provides a theoretical review of the concepts of water asymmetry, inequality and heterogeneity. Section 6.4 describes the research methods and the case study catchment. Focusing on one sub-catchment of Makanya, Section 6.5 presents water sharing arrangements: (a) in one farmer-built and operated irrigation canal, locally known as a furrow ("*mfereji*"); and (b) between furrows in one village. Section 6.6 discusses the initial proposition in light of the research findings and formulates an agenda for further research.

6.3   CONCEPTUAL REVIEW: INEQUALITY, WATER ASYMMETRY AND HETEROGENEITY

Substantial research have been carried on the management of common pool resources and it is now widely accepted that local communities do self-organize to overcome management challenges related to resource use (see Wade 1988, Ostrom 1990, Baland and Platteau 1999). However, there is a lack of conceptual clarity on how collective action emerges and is sustained in settings where water asymmetries exist, namely in cases where different users depend on water flowing along a canal or river.

Water asymmetry arising out of unidirectional flow of river or irrigation water adds to coordination challenge over common pool resources. In a study of river committees in the Themi sub-catchment, Tanzania, Komakech & Van der Zaag (2011) found that water asymmetry negatively impacts on collective action and the river committees' effectiveness. It may also increase the level of inequality, as larger land owners are more likely to occupy land in the head end of an irrigation command area. However, a certain level of inequality in the capacity to access and appropriate a resource may promote collective action. Individuals with higher endowments are in some circumstances more willing to meet the costs of initiating collective action as long as they reap higher benefits from the resource (Baland and Platteau 1999). In a study of agricultural cooperatives in Ecuador, Jones (2004) found that wealthy individuals were more likely to take the entrepreneurial role and initiate collective actions. But he also cautioned that such inequality and exclusive trust may later negatively affect the success of cooperation because of increased free-riding on the wealthy (Jones 2004). Other studies, however, have found that inequality may in fact be a deterrent to cooperation (Varughese and Ostrom 2001), as it leads to unequal sharing of decision-making rights, a low level of trust and unequal allocation of benefits (Bardhan and Dayton-Johnson 2002). Bardhan (2000) found that landholding inequality correlates negatively with maintenance of irrigation canals. He argued that a more egalitarian agrarian society would be more likely to solve collective action problems related to irrigation management. Molinas (1998) posited a nonlinear relationship between inequality and collective action: very low and high levels of inequalities impede collective action, and medium levels are likely to enhance collective action (Molinas 1998). Through large-scale studies of irrigation systems in India and Mexico, Bardhan & Dayton-Johnson (2002) concluded that wealth inequality has a U-shaped relationship with collective action. Low inequality implies that few people have the capacity or can afford to meet the cost of initiating collective action while high levels of inequality may generate resentments (Andersson and Agrawal 2011).

It is evident that there is no consensus in the literature with respect to the relationship between inequality in a resource endowment and collective action. This may be explained because different types of natural resources have been expected to behave similarly, while they do not. We propose that it is useful to distinguish common pool resources with respect to their symmetry, and that fisheries and groundwater fundamentally differ in this respect from water flowing through a canal or river: the effect of one unit appropriated by one fisherperson or groundwater user is the same for all other appropriators and does not depend on their wealth or size or position; for appropriators of flowing water, their impact on other users does depend on their position (whether they are located upstream or downstream). This therefore generates different conditions for collective action.

The question, then, is why does collective action nevertheless emerge among unequal water users? Our hypothesis is that water asymmetry can at least partly be overcome by interdependence, and that interdependence can emerge out of heterogeneity.

Heterogeneity is not well described in the literature on common pool resources and collective action. In many cases heterogeneity is conflated with inequality. This, in our view, is problematic: whereas heterogeneity can be understood as "not being of the same type", i.e. being different (and measured at different metrics), inequality means scoring differently in the same metric. We argue that distinguishing between inequality and heterogeneity may provide conceptual clarity and yield a better understanding of their linkages.

In the literature, some attention has been paid to political heterogeneity, socio-cultural differences and kinship relations. Political heterogeneity relates to the agreement about who is responsible for creating, maintaining and enforcing rights and rules for the use of common pool resources (Vedeld 2000). This dimension is about the leadership roles and authority of individual users in decision-making positions and their legitimacy. Economic elites are more likely to bear the cost of initiating and performing regulatory tasks than others because they are likely to benefit both socially and materially from collective action (Baland and Platteau 1996, as cited in Vedekd, 2000 , Jones 2004).

According to Adhikari & Lovett (2006), ethnic heterogeneity is likely to increase the cost of collective action, because of the need to reconcile a diversity of values among different groupings. Another dimension of socio-cultural heterogeneity is gender. Women's exclusion in decision making may negatively affect collective action (Meinzen-Dick and Zwarteveen 1998, Agrawal 2001), and there is some evidence that increased female representation in decision-making improves the performance of collective action institutions, for example in domestic water supply (Adhikari and Lovett 2006). More importantly, a higher level of female participation does not necessarily mean that the benefits are shared more equitably. Women often get fewer benefits from collective activities than men do.

The literature thus provides some interesting contributions to what could be called "socially constructed heterogeneity," including political, ethnic and gender heterogeneity. Another type is biophysically-induced heterogeneity, which in canal and river systems is primarily linked to landscape features, geology and climate; here diversity is associated with different positions in this landscape. The upper parts may typically have soils derived from parent material (bedrock) with low water holding capacity and a wetter (e.g. sub-humid) climate, whereas the lower parts may have alluvial soils and a dryer (e.g. semi-arid) climate. This biophysical heterogeneity creates different ecological niches in different parts of the command or catchment area, resulting in the production of different environmental goods and services, and conditions the opportunities to produce different crops in different places (for a discussion of farming systems research, see Ruthenberg 1980).

Lansing & Miller (2005) described the role of biophysical interdependencies in promoting cooperative practices among Balinese rice farmers in Indonesia. They observed and explain, using a model based on game-theory, that the selfish behaviour of upstream farmers with respect to water use is mitigated by the threat of crop pests. In the catchment downstream, farmers are more concerned with water shortages while those upstream are concerned about the threat of pests. Without coordinated cropping patterns (e.g. having an identical fallow period) and fair water sharing, everyone is left worse off. It is the realisation of the need for coordination, as dictated by ecological linkages that sustains the farming system (Lansing and Miller 2005).

In this chapter we explore how heterogeneity and inequality play out empirically in situations with unidirectional flows of water. We analyse the canal institutions in the small Makanya catchment area, Tanzania, with the aim of sharpening our understanding of the emergence of enduring water institutions.

6.4   RESEARCH METHODS AND CASE STUDY

6.4.1   Research methods

The research strategy was inspired by the 'follow the water' approach (Latour 1988, Murdoch 1998, Kortelainen 1999, Bolding 2004, Latour 2005). To identify water users and their networks, we followed the water and, in the process, mapped all infrastructures and institutions around it. Fieldwork was undertaken between July 2007 and August 2008 with subsequent visits in 2009. The methods used included mapping, field observation and semi-structured interviews.

We first mapped all six irrigation canals in Bangalala village that rely on water from the Vudee river, using a GPS. Field observation and interviews provided first-hand accounts of irrigation practices. We observed that many farmers had plots along more than one irrigation canal. We then focused on two irrigation canals located on either side of the Vudee river. For these canals we generated a detailed land ownership map. To do this we worked with the elected branch canal (or irrigation zone) representatives, who assisted with mapping the irrigation canals and identifying the owners of the irrigated plots. The land mapping provided us with the opportunity to trace kinship (actors' clans, inter-marriages, etc). We also observed variations in soil characteristics, notably differences in the water holding capacity of soils found in the command area (upstream and downstream).

Next we focused on one of the two irrigation canals, the Mkanyeni canal, and interviewed 31 out of the 83 active members (age 18–79 years; 20 male and 11 female). We collected: historical narratives on the irrigation canal construction; dates for when a farmer joined the water using group and why; land ownership details; crops normally grown; current water allocation rules; details of water variability and how this affects the member's participation; and details of membership of other groups. The interviews made it possible to trace the development of the irrigation infrastructure over time, to see who the founders were, how ownership and membership are defined and how the system is being maintained. During the interviews with farmers, land ownership within the command area of an irrigation canal emerged as a precondition for membership in a particular irrigation group. Membership gives a farmer access to canal water for irrigating crops, but the use of canal water for drinking, washing and for livestock is allowed for all inhabitants of the village. Using the map, we obtained information on kinship and membership of clans and were able to relate this to land ownership. This was done by presenting our detailed land map to the farmers at the irrigation groups' weekly meetings, asking them to clarify the names, and to add information on clans and other related attributes to the map.

We observed weekly water allocation meetings and general elections of new water committees, which provided first hand observations of the dynamics of water allocation but also created opportunities to interview more farmers. We observed that women members complain bitterly about unfair treatment with respect to water allocation. We later interviewed some female farmers and learned that many of them resort to other means to access irrigation water: they may borrow water from the allocation of a relative or connive with branch canal representatives to open the storage dam ("Ndiva") and use the night storage to irrigate at night, often leading to conflict with the farmers scheduled to irrigate the next day.

The irrigation infrastructure was recently improved by NGOs working on food security and value addition. The Same Agricultural Improvement Project (SAIPRO) Trust and Vredeseilanden Country Office (VECO) contributed towards the rehabilitation of the irrigation canals, but also set new conditions of membership and ownership for them. We interviewed the director of SAIPRO about NGO work in the area. We also consulted secondary materials, such as records of attendance at routine maintenance events and the water allocation diaries kept by the irrigation zone representatives. The latter provided information on the number of farmers requesting water, allocated water and the number that actually irrigated. During the research period, translation from Swahili to English and vice versa was undertaken by a native speaker.

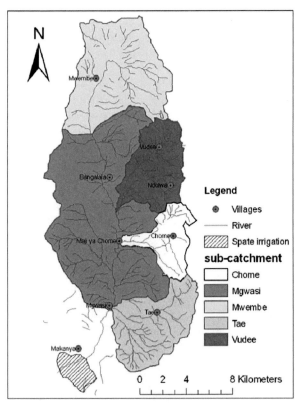

Figure 6.1: Map of Makanya catchment.

6.4.2 Biophysical and socio-economic context

Makanya catchment (300 km^2) is located in the south Pare Mountains in the mid-reaches of the Pangani river basin in eastern Tanzania (Figure 6.1). It lies between latitudes 4°15' to 4°21' S and longitudes 37°48' to 37°53' E, with altitude ranging substantially from 500 to 2,000 m.

The catchment represents typical semi-arid to dry sub-humid rain-fed agrarian conditions, and manifests strong signs of human induced land degradation caused by high pressures on soil and water resources (Enfors and Gordon 2007). Rainfall distribution in the catchment is bimodal (i.e. with a short and a long rainy season). The short season (locally called "Vuli") occurs between October and December, whilst the long season (locally called "Masika") is between March and May. However, in some years, the "short" season lasts longer than the "long" season. Precipitation varies: in the highlands, with 100–500 mm/season and 200–800 mm/season during Masika and Vuli, respectively; in the midlands, between 0–400 mm/season and 50–800 mm/season during Masika and Vuli, respectively; and, in the lowlands, rainfall drops to between 50–300 mm/season and 0–100 mm/season during Masika and Vuli, respectively.

The catchment is drained by the Makanya river, starting from the Shengena Mountain (with a peak at 2001 m above sea level). The river is reported to have been perennial up to the 1970s but has become ephemeral and only reaches the Pangani river during flood events (Mul et al. 2011). Farmers have constructed several furrows across the catchment. In the highlands, apart from domestic and livestock uses, the river water is diverted for dry season irrigation. In the midlands it is used for supplementary irrigation during both rainy seasons, while in the lowlands flash floods are used for spate irrigation during and immediately after flood events (Komakech et al. 2011a). To boost flow into the furrows, highlands and midlands farmers use micro-dams (locally called "Ndivas") to store water at night. Currently over 75 such storage infrastructures exist in the catchment (Mul et al. 2011). Furrow water serves multiple purposes: as well as being used for (supplementary) irrigation, it is also used to meet domestic, livestock, building material and tree nursery water demands. Water sharing practices have emerged to resolve water scarcity-induced conflict in the different parts of the catchment. One such arrangement exists between farmers in the villages of Bangalala, Vudee and Ndolwa located in the Vudee sub-catchment (Mul et al. 2011). Bangalala is in the lowlands, while Vudee and Ndolwa are in the upstream part of the sub-catchment.

This case study analyses the nested water sharing practices that have emerged between farmers in Bangalala village on two scales. In 2008, the population of Bangalala was estimated to be 3800 (Same District report 2008). Bangalala farmers use Vudee water for irrigation and have developed six irrigation canals. The farmers practice supplementary irrigation during the rain seasons (Masika and Vuli) and full irrigation in the dry period (June–September and January–February). Water shortages are severe in the period July–October and the goal of most farmers is to secure long-term access to water for irrigation, domestic and livestock needs. In the next section we first describe the water sharing arrangements within one irrigation canal (Mkanyeni) and then the arrangements between canals in one village (Bangalala).

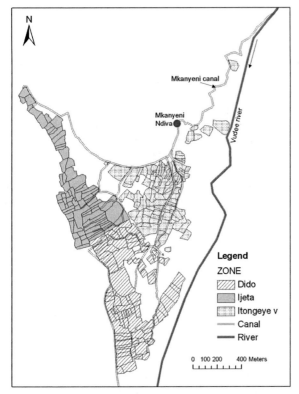

Figure 6.2: Mkanyeni irrigation canal, micro-dam (Ndiva), zones and plots.

6.5   WATER SHARING ARRANGEMENTS

6.5.1   Evolution of water sharing practices in Mkanyeni furrow

The Mkanyeni furrow (Figure 6.2) has a command area of about 94.8 ha, and consists of 1.6 km of primary canal, a micro-dam (storage capacity of approximately 1,015 m^3), and two major secondary canals. The canals are unlined, while the intake is made from stones, twigs and mud (it leaks and regularly gets destroyed by floods). Significant labour efforts are required to sustain the furrow system. The intake capacity is not fixed (in May 2008 the average flow in the main canal at the intake fluctuated between 40 and 56 l/s). The secondary canals form a canal network of about 9 km serving three irrigation zones, each having a name (Dido, Ijeta and

Itongoye vivi; see Figure 6.2). Itongoye vivi is upstream and, during low flows, water allocation is restricted to this zone. Ijeta and Dido are in the flood plain and the soils have a higher water holding capacity compared to soils in the upstream areas. In 2008 there were 83 members (membership has varied in the past, being 90 in 2006 and 79 in 2007). Farmers can leave the group as and when they deem fit but they are reportedly charged a fee to re-join of Tanzania shillings (Tsh.) 10,000 (equivalent to 8.40 USD at the time of fieldwork in 2008).

It is not clear when the furrow ("mfereji") was constructed, but farmers interviewed say the Kitojo and Safieli Chungankwi families constructed it in about the early 1940s. When other families whose plots could be irrigated requested to join, they were allowed to become members. New entrants would provide a kilo of sugar or five litres of local alcoholic brew as an entry fee to the leaders of the furrow founding families and had to commit to participate in the maintenance of the canal. From 2001, new members have paid a one-off membership fee (called "Ukarabati", which is also translated as a rehabilitation fee) of Tsh. 500 (0.42 USD), and a registration fee (called "Kiingilio", translated as "entrance") of Tsh. 500 (0.42 USD). The membership fee is meant as a contribution towards the prior investments made by the group.

To capture night flows and avoid irrigating at night, the families constructed a micro-dam ("Ndiva") using stones and clay in about 1949. In 1952, the management of the furrow system was given to Lungiro Kitojo, the son of Kitojo, while Safieli Chungankwi constructed another micro-dam downstream (according to farmers interviewed, Chungankwi was a famous mason in the village and considered very innovative). Lungiro allowed more people to join but his descendants later wanted to regain complete control of the furrow system, claiming it was their inheritance. The village elders, however, advised them to share the furrows with the other members since they contributed to maintaining it. Descendants of the founding families still have a location advantage, i.e. they own plots in the upstream parts of the command area, and they also own the biggest plots in the irrigation zones. In the 1970s, the government abolished natural resources management by clanship; as a consequence the irrigation system was declared village property but they continued to be managed by the founding families. By 1978, membership had increased beyond the capacity of the system, so the group decided to enlarge the micro-dam in order to address the increase in water demand and store more water for its members. They also lined it using cement to reduce water leakage. By 1999, the micro-dam was again leaking and the group sought assistance from non-governmental organisations and CARITAS (an international NGO working in Same District) provided cement and food for work during construction, whilst VECO (another international NGO working in Same District) provided a bulldozer for the micro-dam expansion. In 2002, the Ndiva ya Mkanyeni (abbreviated to Ndimka) water user group was formally established.

In 2003, the micro-dam again started leaking and the farmers sought assistance from SAIPRO. SAIPRO agreed to provide resources for its reconstruction but on condition that the farmers create and adopt a written constitution, that they pay a so-called "commitment fee" of Tsh. 5,000 (4.20 USD) each to SAIPRO, and that they all contribute by gathering stones, paying for the mason, pipes and gate valve. To meet this latter requirement a seasonal fee (called "Ada") of Tsh. 1,000 (0.84 USD) was introduced. In 2007, the Ndimka group started the process of developing a written constitution ("Katiba ya Ndimka"). The constitution, which defines roles and responsibilities of members and sanctions, includes old and new rules. Box 6.1 gives an extract of principles from the draft constitution. Note that, in the draft constitution, some of the old norms and traditions have been changed in terms of gender roles, age and membership. For instance, membership and registration fees were increased, a new rule acquitting older members from maintenance activities was introduced, and anyone with access to land could become member and a leader (this rule allows female members the opportunity to become leaders, something that was not allowed in the past). Money collected by the irrigation group is used to procure materials such as cement, tools (e.g. spades) and for paying masons during repairs of the Ndiva.

Box 6.1: Extracts from Ndimka draft constitution. (Source: translated from Swahili to English by the authors).

- Anyone above 18 years of age who lives in the village and owns a plot within the command area can be a member;
- All new members pay an entry fee of Tsh. 1000 and a registration fee of Tsh. 5000 but members with salaried jobs pay Tsh. 2000 as an entry fee;
- All farmers pay a seasonal (per half year) fee of Tsh. 1000;
- Members who are 70 years of age or older are excused from working; disabled persons pay a nominal fee;
- Members in formal employment, such as teachers and other government employees, are excused from maintenance activities but must contribute Tsh. 10,000 yearly;
- Not participating in maintenance activities, delay in paying seasonal fees or missing meetings attracts a penalty of 2000 Tanzanian shillings and defaulters must still do the work after paying;
- Any member caught stealing water will pay Tsh. 10,000, and is given a warning. If caught three times, he or she will be dismissed from the group. Non-members caught stealing water are to be sued through the formal legal procedures; and
- Bathing, washing clothes and watering animals inside the canal is prohibited, and a fine of Tsh. 10,000 applies, because it pollutes and damages the canal banks.

6.5.2   Furrow management and sustainability

Although all furrows were declared village government property in the 1970s, their management has remained outside local government structures. Instead of the founding families providing leadership, a periodically elected water committee is now in charge of Mkanyeni. The committee is responsible for water allocation, conflict resolution, supervision of maintenance activities, monitoring compliance, calling for meetings and negotiating water sharing arrangements with the other furrow groups in

Bangalala village, as well as in the other upstream villages. The committee is comprised of a chairperson, vice chairperson, secretary (who also doubles as a treasurer), four elders ("Wazee washauri", translated as council of "wise" elders), and three irrigation zone representatives ("Halmashauri"). Since 2001, elections for the committee have been held every three years. The election process is supervised by the village government, and this is the only major role the village government has in furrow management. We witnessed the elections of Mkanyeni and Manoo water committees in April and May 2007, respectively. The positions of chairperson, secretary and irrigation zone representatives are elected by secret voting (using a ballot box), while that of the elders is by consensus. The position of vice chairperson is filled by the runner up for the chairperson position. However, for both Manoo and Mkanyeni, one week before the election day, the outgoing chairperson and his committee nominated three persons whom they considered fit to fill the chairperson position. If the sitting chairperson is still interested, his/her name could have been included (this was the case for Mkanyeni and the sitting chairperson indeed got re-elected). The names of these individuals were floated to the members for election and members voted secretly under the supervision of the village government representative. For the zone representatives ("Halmashauri"), names were proposed by farmers from the respective zones and votes cast to elect the representative for each canal. For the elders (the Wazee Washauri), four individuals were also nominated by the farmers. Each irrigation zone nominated and elected their elder, and the presiding officer from the village government conducted the vote once the farmers present at the election had nominated the candidates. The criteria for nomination of the elders were that all three zones must be represented and that at least one of the elders must be female. The elected committee then elected the secretary/treasurer.

Although we witnessed that the election process was transparent and fair, through spatial mapping of the furrow system we found that the elected individuals or their relatives tend to have access to the largest irrigated area. In addition, those elected for the chairpersons' positions were all descendants of the irrigation canal founding families. By preselecting the candidates for chairperson, the founding families continue to maintain some control over the furrow system. Farmers with many plots frequently lend some of their plots to friends, which may increase their chances of being selected as leaders.

6.5.3   Land access inequality and heterogeneity

We mapped all the plots within the irrigation area and found that 132 farmers (37 of them female) have plots in the command area, with average land access of 0.69 ha. Only 79 (27 female) of the plot owners are currently active member of the irrigation group. Four of the active member borrowed their land. Figure 6.3 shows that the top 20% of farmers (in terms of largest land access) own 50% of the irrigation area and that the bottom 50% of farmers (in terms of smallest land access) control about 20% of the land area. In addition to having land in all three irrigation zones, farmers belonging to the founding clans (Mmbaga and Mshana) also control the largest land area in the Mkanyeni system (Table 6.1). The Gini coefficient of land access in Mkanyeni is 0.58[12]. Moreover, nearly half of all farmers (49%, 65 out of 132) own at least two plots in the command area and one-third (46 out of 132) have plots in at least two zones. The latter fact evens out, to some extent, the impact of upstream priority allocation during periods of water shortage.

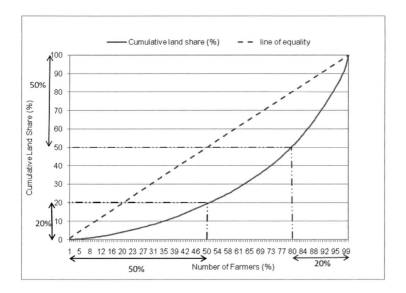

Figure 6.3: Unequal land access in Mkanyeni furrow system: Top 20% of the farmers (in terms of largest irrigation land access) own 50% of the irrigated land and the smallest 50% of the farmers (in terms access to the smallest land area) control only 20% of the irrigation land area (source: field mapping).

---

[12] The Gini coefficient is calculated as the ratio of the area between the line of equality and the Lorenz curve (the line of cumulative land share in Figure 6.3) and the total triangular area under the line of equality.

Table 6.1: Land access distribution by clan in the three zones of Mkanyeni furrow.

Clan	Dido (%) Area: 38.6 ha	Ijeta (%) Area: 27.5 ha	Itongoye (%) Area: 28.7 ha	Total (%) Area: 94.8 ha
Mchaga	0.5	0.5	0.2	0.4
Mchome	2.2	-	7.2	3.1
Mgiriama	1.2	2.8	0.5	1.4
Mgonja	1.5	6.9	-	2.6
Mjingo	0.7	11.9	1.3	4.2
Mkiramweni	2.4	0.9	16.0	6.1
Mmbaga	13.0	6.6	19.2	13.0
Mmbugu	0.8	-	-	0.3
Mnyone	3.2	2.5	4.8	3.5
Mrutu	9.0	14.6	5.6	9.6
Msambaa	1.3	-	3.3	1.5
Mshana	49.9	33.3	17.2	35.2
Mtaita	0.3	5.9	3.0	2.7
Mtango	1.9	-	3.0	1.7
Mngulu	-	1.8	-	0.5
Mdoe	-	-	2.3	0.7
Msemo	-	-	1.8	0.5
Unknown	12.1	12.3	14.6	12.9
Total	100.0	100.0	100.0	100.0

Furthermore, the location advantage enjoyed by upstream farmers (Makurira et al. 2007, Kemerink et al. 2009) seems to have little impact on crop yields. This is true for both Mkanyeni and Manoo furrows. In semi-arid areas where water is a limiting factor, the water holding capacity of the soil plays a significant role. Bangalala furrow system farmers all have their downstream plots located in the flood plain where weathered alluvial materials are deposited (Mul et al. 2007a). Downstream plots therefore have soils with higher water holding capacity compared with upstream plots. The farmers are aware of this added advantage and during interviews they reported that one irrigation turn or water allocation to plots in the downstream area is sometimes sufficient to obtain a good harvest.

6.5.4   Furrow water allocation, conflict and gender

The Ndimka water committee meets every Thursday, normally at 2pm, with the group members deciding on the water allocation for the coming week, discussing maintenance issues, and to solve conflict, among others. Although Ndimka's draft constitution requires all members to attend, only farmers needing water are normally present. As further described below, Mkanyeni furrow gets water for three or four days per week. Within the furrow, water is first allocated to the irrigation zones (Dido, Ijeta and Itongoye vivi) and then to the farmers in each zone. Water allocation between the irrigation zones within Mkanyeni furrow is made in turns: each zone gets a one day allocation per week, except in the week that Mkanyeni takes river water for four days when one of the zones is allocated two days; this additional day is rotated amongst the zones every time it occurs. To minimise losses during extreme shortages, water can be allocated to the upstream zone only. Itongoye vivi (located nearest to the micro-dam) was observed to receive more irrigation turns during the dry season than the other zones.

Unlike water allocation between the zones, which is directed by the group's chairperson, water allocation to individual farmers is conducted with direct assistance from zone representatives and the council of elders ("Wazee Washauri"). Water allocation to a farmer depends on the zone, water availability, the state of the crop and the condition of the distribution canal. Allocation is per farmer and not time-based. Farmers with plots in more than one zone are allowed to irrigate only one plot when there is insufficient water. On average, three farmers per day can be allocated water in the system and some of the water allocation criteria include: whether the farmer prepared the canal section leading to the plot, participated in routine work, or paid the membership fees. It is the responsibility of the zone representative to ensure that farmers in his/her zone fulfil these conditions. However, they do not rigidly adhere to the requirements; there are situations where the canals are cleaned after allocation has been made.

When the river carries sufficient water, allocation usually starts with downstream zones or plots, moving upstream, but during periods of low flow, water allocation starts upstream. Sometimes the furrow committee anticipates excess flow (referred to as "mafuriko", the Swahili for flood) and allocate water to six or seven farmers (more than the average three farmers per day). If indeed there is excess water, additional farmers can also irrigate, otherwise they have to wait for their own turn or are given priority in the following week's allocation. Once the week's allocation schedule is ready, it is the responsibility of the farmers who have been allocated water to reconstruct the main furrow intake, clean the main canal leading to the micro-dam, operate the micro-dam and ensure that water transmission losses are minimised.

The weekly water allocation meetings can be very chaotic, and emotions can build. Scenes in which everybody stands up, and of women complaining bitterly about unfair treatment, are common. All eight meetings we attended between November 2007 and May 2008 demonstrated high emotion. On 27 May 2008, while undertaking spatial mapping in Manoo furrow, we heard the voices of women 3 km away at the water allocation venue. The women were complaining that the water committee was allowing some farmers to get a second irrigation turn, when they had not yet received their first. The women shouted for about 3 hours but, when the meeting ended, their voices were ignored. The following day, we found that some of these women had woken at 3am and taken another farmer's allocation to irrigate their plots. This resulted in quarrels the next morning which were later solved at a personal level, the women admitting guilt but refusing to compensate the offended person.

Furrows are often cleaned at the end of the Masika season in May to prepare for full irrigation during the dry season. Each farmer (member) is allocated a section to clean/repair (normally 17–20 wide steps) by the representatives. Cleaning the micro-dams is also undertaken jointly but the maintenance of tertiary canals is the responsibility of the members served by them. After repairing collapsed sections of the canals, there is less leakage along the furrow network. During our field work in May 2008 (i.e. at the end of the Masika season), desilting of the Mkanyeni micro-dam took 7 days, while cleaning the transmission canal took two weeks, with about 45 farmers (the majority of whom were men) being present each day. The group stated that one week is enough for canal cleaning unless interrupted by unforeseen social events (burials, weddings and NGO meetings were indeed frequent during the period February–June 2008).

In practice, participation in furrow maintenance activities and the irrigation requirements of one's crops are not the only criteria to qualify for an irrigation turn. Farmers use other means to access water including, amongst others, utilising social relations with the furrow leaders (such as friendship and family ties), or using personal standing in the village (some of the vocal farmers were retired government civil servants, retired police officers and military officers). Farmers also share plots near the micro-dam with fellow farmers during times of extreme water shortage. A farmer upstream and near to the dam loans his/her plot or part of it to a friend downstream and, at the discretion of the downstream farmer, the plot owner may receive some gifts in return (e.g. salt or a portion of the harvest, a procedure known in Swahili as "zawardi"). We followed one farmer who shared some of his plots with others. This farmer is a descendant of one of the founding families and inherited a large land area from his father, who also inherited it from his father. The farmer is the elected representative of 35 farmers in the Ijeta zone, a position he has held for four years now. He is currently doing agri-business in another district and not able to farm all his land. He therefore delegated his zone representative role to another farmer, and shares most of his large land area with other farmers. He was subsequently re-elected as zone representative. The land-lending/gift system in this case seems to have been important for gaining and maintaining a leadership position.

Thus, the Mkanyeni case provides insight into the interaction between inequalities (e.g. access to land) and heterogeneity (e.g. differences in soils between the zones, farmers having many plots and kinship). The interaction configures power relations as seen in the attempt by founding families to maintain control over leadership roles by preselecting candidates for the chairperson's position and in women's continued exclusion in spite of their protests. The emergent power relation is constantly being contested and reinterpreted. While these collective arrangements seem to work for men, women are clearly disadvantaged. Dynamic interactions are also embedded within the wider biophysical and socio-economic context of the Makanya catchment. The water sharing arrangement in Mkanyeni furrow influences and is also influenced by other furrows both in Bangalala village and further upstream (e.g. furrows in Vudee and Ndolwa villages). In the following section, we make the water use connection between Mkanyeni and three other furrows in Bangalala village and show how interdependencies emerge to counterbalance water asymmetry at this scale.

6.5.5   Water sharing between furrows in Bangalala village

Currently four major furrows in Bangalala (namely Mghungani, Kinyanga, Manoo and Mkanyeni) use Vudee water for irrigation, domestic and livestock purposes. Mghungani and Kinyanga share one intake structure on the Vudee river before its confluence with the Ndolwa tributary and the section of the conveyance canal between the intake and the Kinyanga micro-dam (the distribution networks of the two furrows are not shown in Figure 6.4). Mkanyeni and Manoo only share the intake point, while minor furrows such as Mondo wangombe and Chungangwi are considered to be part of Manoo furrow (Figure 6.4).

Water allocation at the village level is first made by tributary. The other two (Manoo and Mkanyeni) abstract water downstream of the confluence but this water is considered to come from the Ndolwa tributary. This arrangement was arrived at based on the historical development of the furrow systems. Mghungani is said to be the oldest furrow in the village, followed by Manoo, then Mkanyeni and Kinyanga. Since Mghungani was already abstracting water from the Vudee tributary, Manoo furrow was allocated water from the Ndolwa tributary. It is further important to note that the intake structure for Mghungani and Kinyanga is made of concrete and can only abstract about 50 l/s, leaving the excess water to flow downstream.

Water allocation between furrows sharing the same intake point is made on a rotational basis. For Mghungani and Kinyanga furrows, allocation is based on the estimated command area, with the former taking water for 5 days (since it is said to serve a bigger area) and the latter taking water for 2 days. However, weekly allocation for Manoo and Mkanyeni is based on a 4-3-3-4 system (i.e. in a week, one furrow abstracts water for four days and the other three; the following week, the first furrow has three and the other has four days). The water allocation schedule between the Manoo and Mkanyeni also takes into account religious holidays. The furrow shares the inconvenience of working on a religious holiday by allowing one furrow to divert water on Friday (Muslim prayer day) and the other on Sunday (Christian prayer day), while the Saturday, which is the prayer day for Seventh Day Christians, alternates between the two furrows. According to the farmers, the schedule was designed by the village elders to avoid religious conflict.

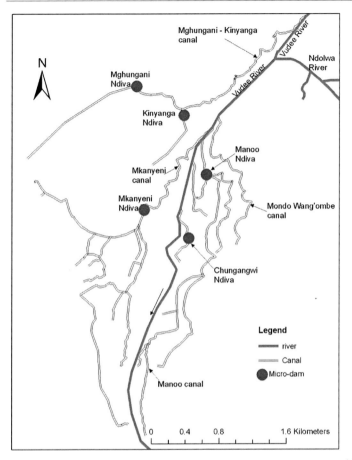

Figure 6.4: Bangalala village furrow system and micro-dams (locally called Ndiva).

The mapping of access to irrigated land in the Mkanyeni and Manoo furrow systems provides further insight to the water sharing arrangements between the two furrows. The land map shows that most clans own land in both systems (Figure 6.5). Interestingly, no clan is dominant in both furrows. The table in Figure 6.5 shows that the Mnyone clan controls just over 3.3 hectares in Mkanyeni and as much as 43 hectares in neighbouring Manoo furrow; in contrast, the Mshana clan controls 33 hectares in Mkanyeni and only 13 hectares in Manoo.

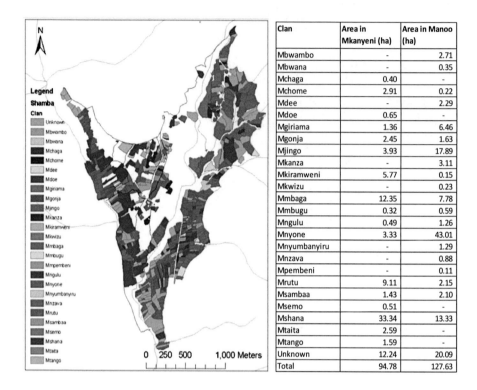

Clan	Area in Mkanyeni (ha)	Area in Manoo (ha)
Mbwambo	-	2.71
Mbwana	-	0.35
Mchaga	0.40	-
Mchome	2.91	0.22
Mdee	-	2.29
Mdoe	0.65	-
Mgiriama	1.36	6.46
Mgonja	2.45	1.63
Mjingo	3.93	17.89
Mkanza	-	3.11
Mkiramweni	5.77	0.15
Mkwizu	-	0.23
Mmbaga	12.35	7.78
Mmbugu	0.32	0.59
Mngulu	0.49	1.26
Mnyone	3.33	43.01
Mnyumbanyiru	-	1.29
Mnzava	-	0.88
Mpembeni	-	0.11
Mrutu	9.11	2.15
Msambaa	1.43	2.10
Msemo	0.51	-
Mshana	33.34	13.33
Mtaita	2.59	-
Mtango	1.59	-
Unknown	12.24	20.09
Total	94.78	127.63

Figure 6.5: Land control by major clans in Mkanyeni and Manoo furrow systems (source: field notes).

It is evident from the heterogeneous pattern of land ownership that any member of a clan controlling the largest share of irrigated land in one furrow cannot just ignore the interests of the other clans, as any selfish act may be reciprocated in another furrow where the clan is not dominant. Here inequality in access to land (in terms of some clans controlling more land than others) again combines with heterogeneity (farmers having plots in many furrows and multiple membership) to mitigate the negative impact of water asymmetry between furrows in Bangalala village.

6.6   DISCUSSION AND CONCLUSIONS

This chapter has aspired to provide a conceptual clarity to situations where collective action has emerged and has been sustained despite water asymmetry. We first studied internal arrangements with respect to water sharing within one canal organisation in Makanya catchment, Tanzania, and found a certain level of inequality of access to land (Gini coefficient of 0.58) and thus of access to water. This inequality might be understood as the outcome of underlying dynamics that limit excesses: if land and water become concentrated in too few hands, irrigators may opt out and shift their efforts to another canal, making the mobilization of labour for maintenance more problematic, and hence possibly leading to system collapse. This hypothesis could be verified by measuring the inequality of access to land and water in many furrow systems, and by investigating any variation found in relation to the collective ability to share water and mobilize labour for maintenance.

But this inequality alone appears insufficient to explain how these canal organisations have been able to endure. We consider heterogeneity to be an additional factor that has led to mutual dependencies among the Bangalala village irrigators. First, there are systematic differences in soils in the upper and lower parts of the command areas of the canals, which not only leads to differences in water requirements and frequency of watering, but also to different crops being grown (e.g. vegetables and maize in the upper and lower part of the command areas, respectively); irrigators need both, and can barter and trade. Second, water users along one canal have close kinship ties with other irrigators in the command area of the same canal, and in neighbouring canals. Third, upstream irrigators cannot completely ignore downstream counterparts, due to the large labour requirements needed to repair and maintain the river intake, the main canal and often also the storage tank. Fourth, many of the irrigators have plots in different parts of the command area, resulting in irrigators having multiple interests and loyalties, which inhibits unilateral action.

The interactions between water asymmetry, inequality and heterogeneity influence and configure power relations and hence the nature of water allocation. We observed that inequality in access to irrigated land also leads to disproportionate water allocation. It is farmers with larger land endowments, with many family members among the group of irrigators, and with good communication skills, who are able to secure access to more frequent water turns to irrigate their larger area or their several plots located in different zones. We observed in Mkanyeni's Thursday water allocation meetings that farmers with more land tend to dominate the meetings, and that decisions often go their way. Women tend to lose out, although they often resort to alternative ways to access water (e.g. "theft", borrowing from neighbours, etc). The larger landowners, who in most cases are members of furrow founding families, continue to maintain control by preselecting leaders of the canal organisation. They continue to influence furrow management to this day even when the government abolished clan-based water management in the 1970s. Apparently, this degree of inequality does not jeopardise the integrity of the system, an observation which supports the findings of Molinas (1998) and Bardhan & Dayton-Johnson (2002).

The transformation over time and the "irregularities" of water property rights observed in Makanya catchment have to be discussed within the context of plural institutions, plural laws and contested ownership found in Tanzania. All water resources were first declared public property by the colonial powers as early as 1923. The Water Ordinance of 1948, chapter 257 (Territory 1948), stipulated in section 4 that "the entire property in water within the Territory is hereby vested in the Governor, in trust for His Majesty as Administering Authority for Tanganyika [...]." They also set the condition for the development of plural institutions by allowing areas under "natives" to be governed by local customs and traditions and those areas declared crown land to be managed by state-led water laws. The independent government subsequently abolished certain customary practices and opened up ("villagised") access to irrigation water to all villagers. However, old practices and powers persist and, in Makanya, irrigation water has never become fully public. The exclusion, conditions of entry and opportunistic behaviour (entry fees, limiting women's access, etc.) observed in the furrows show that irrigation water might better be considered a club good or toll good (cf. Buchanan 1965). This can be explained by the fact that the infrastructure has a history, and that the process of hydraulic property creation (Coward 1986b) affects, and is affected by, power relations. Physical structures and social relationships thus co-evolve. Their specific combination as found locally is thus not accidental, and mediates water access and control in the catchment.

The findings presented in this chapter are relevant to our understanding of the functioning of water institutions more generally in three ways. First, inequality among users is not naturally a deterrent to collective action, and may even be considered a resource that can help initiate and maintain the collective good. However, collective action may still mirror other inequalities, as is the case with gender. Second, mutual dependencies that exist between users and user groups in a water system may similarly be considered to be a vital resource, and making such dependencies explicit or even consciously increasing them helps to stimulate collective action. Third, the combination of inequality and interdependencies may give rise to emerging dynamics that can explain sustained collective action in situations of water asymmetry. This is exemplified by the fact that the canal organisations described have been functioning for several generations, and have not collapsed but instead are still functional.

The current findings are confined to relatively small spatial and social scales, involving irrigators from one village. In such situations there may be inhibitions to unilateral action due to social and peer pressure. Spatial and social proximity may thus be a necessary condition for collective action in water asymmetrical situations to emerge. At larger spatial scales and over greater distances, for example when considering catchment areas or river basins, this is likely to be different. The social relationships that could promote collective action in such larger spatial scales have hardly been studied but could include inter-village marriages, church groups, seasonal or longer-term migration patterns, pastoral movements, as well as formal representation in local and district and higher-level government. A better understanding of such relationships, their inequalities and heterogeneity, may help identify already existing incentives for collective action when need arises. Those characteristics could well be powerful complementary arrangements to formal, top-down established basin organizations. Research is therefore needed to describe phenomena of water asymmetry, inequality and heterogeneity at larger spatial scales, and to analyse under which circumstances they occur. Such research could then connect with studies that discuss the possibilities and constraints of issue linking at the transboundary scales (e.g. Fischhendler et al. 2004, Meijerink 2008, Dombrowsky 2010).

Chapter 7

Understanding the Emergence and Functioning of River Committees in a Catchment of the Pangani basin, Tanzania[13]

7.1 Abstract

In this chapter we explore the emergence and functioning of river committees (RCs) in Tanzania, which are local water management structures that allocate and solve water conflict between different water users (smallholder irrigators, large commercial farmers, municipalities, etc) along one river. The chapter is based on empirical research of three committees in the Themi sub-catchment. The committees mostly emerged in response to drought-induced competition and conflict over water, rapid urbanisation around Arusha town, and the presence of markets for agricultural produce. The RCs are mainly active during dry seasons when water is scarce.

We find that the emergence of the RCs can be understood by using the concept of institutional bricolage. We then assess their effective functioning with the help of the eight design principles proposed by Ostrom and find that the best performing RC largely complied with five of them, which indicates that not all principles are necessary for a water institution to be effective and to endure over time. Neither of

---

[13] Based on Komakech and van der Zaag, 2011. *Water Alternatives* 4(2), 197-222

the other two studied RCs complied with three of these principles. All RCs leave the resource boundary open to negotiation, which lowers the transaction cost of controlling the boundaries and also allows future demands to be met in the face of increasing resource variability. All RCs do not fully comply with the principle that all affected must take part in rule creation and modification. In all three cases, finally, the "nesting" of lower-level institutional arrangements within higher-level ones is inconsistent.

To explain the difference in the performance of the three RCs we need to consider factors related to heterogeneity. We find that the functioning of RCs is strongly influenced by group size, spatial distance, heterogeneity of users and uses, and market forces.

7.2 INTRODUCTION

In this chapter, we explore the emergence, functioning and challenges faced by river committees (RCs) in the Themi sub-catchment, Pangani river basin, Tanzania. RCs are locally developed water management structures that manage water allocation and solve water conflicts between groups of water users (e.g. between irrigation canals, locally known as *mifereji* (singular: *mfereji*) or furrows, commercial farmers, municipalities, etc.) of one river but not between individual water users. In the Themi sub-catchment, RCs bring together large- and small-scale users, districts, divisions, wards and village leaders. A committee typically controls only a part of the river watercourse. Several studies have been documented on the evolution and the effectiveness of self-governing water institutions (Fleuret 1985, Grove 1993, Ostrom and Gardner 1993, Ostrom et al. 1994, Potkanski and Adams 1998). Most of these studies looked at dynamics of water rights in a community's traditional irrigation system; few, if any, document institutional arrangements created by the users to manage water allocation between small-scale irrigation schemes, villages and large commercial estates along one river. Understanding such arrangements can add useful knowledge on local institutions, as well as on the establishment of new intermediate water or river institutions. We use three different scientific perspectives to examine how RCs emerged and currently function in different parts of the Themi sub-catchment.

To understand the emergence and functioning of local institutions, we review (a) the concept of institutional bricolage introduced by Cleaver; (b) Ostrom's eight design principles of long-enduring, self-governing institutions; and (c) the role of heterogeneity and group size. Then we present three case studies of RCs in the Themi sub-catchment. We subsequently analyse the emergence and functioning of these RCs with the help of the three sets of concepts. The concept of institutional bricolage helped to explain the emergence of the RCs but does not provide insight into their effectiveness with regard to resource management. Ostrom's eight design principles provided useful entry points in studying the functioning of the RCs, but these were not sufficient to explain how well the RCs regulated resource use. Surprisingly, three of Ostrom's principles were largely or completely absent. Group size, spatial distance, heterogeneity of users and uses, and market forces were found to pose important constraints to the effective functioning of the RCs.

7.3 THEORETICAL FRAMEWORK: INSTITUTIONAL EMERGENCE AND FUNCTIONING

Because of its rivalry and non-excludability characteristics, river water is a common pool resource (CPR) (Ostrom and Gardner 1993). In addition, because of location asymmetry, appropriation of the water resources leads to potentially conflicting situations between upstream and downstream users, in that upstream users can exercise their claim to water first (Ostrom and Gardner 1993, Van der Zaag 2007). Downstream users require the cooperation of their upstream counterparts, which underscores the relevance of collective action over CPR management. However, collective action for CPR management can have various interpretations and may involve a complex set of rules and institutions arising out of historical, ecological and other structural processes (Ostrom 1993, Ostrom 2000, Naidu 2009). In this chapter, we define collective action as the contribution users make to allocate, manage and regulate the use of water resources (e.g. in terms of actors' time in attending meetings to discuss water issues, restraining from using the resource out of turn, joining work parties for canal repair and maintenance, and actively monitoring the resource).

7.3.1   Design principles for long-enduring institutions

Ostrom identified eight general design principles (Table 7.1) for long enduring, self-governing CPR institutions (Ostrom 1993, Ostrom et al. 1999, Ostrom 2000, Dietz et al. 2003). These eight principles are presented as elements strongly correlated with the success of long-enduring institutions in sustaining a particular CPR and gaining the compliance of generation after generation of resource appropriators to the rules-in-use (Ostrom 1993, Ostrom and Gardner 1993, Sarker and Itoh 2001).

Scholars have criticized the theoretical grounding of these design principles or argued that the principles do not offer a comprehensive solution to CPR problems, or that their application may lead to simplistic attempts to force institutions to conform to the principles regardless of relevance or feasibility under particular conditions (Cleaver 2000, Agrawal 2001, Cleaver 2002, Cleaver and Franks 2005, Bruns 2009). Agrawal and Gibson (1999) maintain that particular characteristics of a community, however defined, may not predict outcomes but may rather influence the process of institutional formation. Cleaver and Franks (2005) add that because of complexity, diversity and the ad hoc nature of institutional formation, institutions elude design.

Table 7.1: Brief review of Ostrom's eight general design principles of self-governing CPR institutions.

Design principles	Critical reviews
1. Clearly defined boundaries (of resource and users)	This principle ensures that appropriators can clearly identify anyone who does not have rights and take action against them. However, scholars argue that this requirement is too rigid, as it may fail to account for resource mobility and variability. Also, for some CPRs, the users' boundaries are not "waterproof", as they are dynamic over time (Quinn et al. 2007) and at times may include non-resident users (e.g. distant cities relying on a river for hydroelectricity production).
2. Congruence between appropriation and provision rules and local conditions	This principle refers to rules being considered fair and legitimate by the users and also match local conditions, e.g. soils, slope, number of diversions, crops being grown, etc. This is relatively straightforward in the case of an irrigation canal, where labour for maintenance of division gates and canals mediates allocation turns. At the level of a river appropriation rules specifying individual furrow water entitlements are not related to their inputs (no labour, material, and/ or money invested in resource provision). Provision refers to upstream users' willingness to agree to water-sharing arrangement without being compensated. The dissimilarity between provision and appropriation rules makes enforcement difficult.
3. Collective choice arrangements	Individuals affected by the operational rules can participate in modifying these; otherwise some appropriators may perceive their costs as higher than their benefits and cheat whenever an opportunity arises. Cheating increases enforcement costs (Ostrom 1998). A system where some actors are able to cheat while others conform to the rules is unlikely to survive for long.
4. Monitoring	This principle refers to the presence of monitors who actively audit CPR conditions and appropriator behaviour, and who are accountable to the appropriators or are appropriators themselves. But resource users often get involved in ad hoc monitoring arrangements, e.g. downstream users pay guards to monitor use upstream (Ostrom 1993, Sarker and Itoh 2001).
5. Graduated sanctions	Violators are sanctioned by their peers, and get increasingly severe sanctions if they persist. However, this does not often happen. Some actors use a 'forum shopping' approach (see, for instance Meinzen-Dick and Pradhan 2002) to settle conflicts (e.g. appropriators use various laws and norms to argue and settle cases, some go to the government court, while other cases get solved by traditional authorities, or at a personal level).
6. Conflict resolution mechanisms	Appropriators and their officials have rapid access to low-cost local arenas to resolve conflicts among appropriators or between appropriators and officials.
7. Minimal recognition of rights to organise	The appropriators' rights to devise their own institutions or their legitimacy not being challenged by other authorities.
8. Nested enterprises (hierarchical or interrelated organisation levels)	The principle implies the existence of a direct hierarchical relationship within the group, with other groups, and/or higher-level authorities in a catchment. But local management structures are rarely hierarchical, may be ad hoc, dynamic and frequently renegotiated (Cleaver and Franks, 2005). The principle also claims that nesting each level of organisation within a larger level allows for the externalities that are caused by one group and imposed on others, to be addressed by a higher-level structure. So smaller units can take advantage of economies of scale where they are relevant and to aggregate capital for investment (Ostrom 1993). However, for RCs nesting may not directly translate to no shirking within a furrow. Individual furrow management is largely independent of the working of the RCs.

7.3.2   Heterogeneity and group size

Although there is a rich body of literature on the role of heterogeneity and group size on collective action for common pool resource management institutions no consensus exists (see, for instance Olson 1965, Bardhan and Dayton-Johnson 2002, Poteete and Ostrom 2004, Ruttan 2008, Araral Jr. 2009, Naidu 2009). This section gives a brief overview of the debate, starting with the role of heterogeneity.

Heterogeneity may include sociocultural diversities, wealth inequalities, inequalities in sacrifices members make in cooperating with collective management regimes, locational differences not reflected in landholdings and wealth (e.g. upstream-downstream asymmetry in water access), inequalities in outside earning opportunities (exit options) and benefit heterogeneity (i.e. heterogeneity in economic interests) (Baland and Platteau 1999, Bardhan and Dayton-Johnson 2002, Naidu 2009). While some empirical studies find that sociocultural heterogeneity negatively affects collective actions (Bardhan and Dayton-Johnson 2002) others report that its impact is positive or in some cases insignificant (Poteete and Ostrom 2004). Baland and Platteau (1999) argue that economic inequality under certain conditions leads to higher provision of collective goods. Others have argued that economic inequality is most important in initiating collective action but tends to negatively affect full participation (Varughese and Ostrom 2001, Poteete and Ostrom 2004, Ruttan 2008, Naidu 2009). Some theoretical researchers, however, identify a U-shaped relationship between economic inequality and cooperation, suggesting that both schools of thought may, in part, be correct (Bardhan and Dayton-Johnson 2002).

To resolve the impasse, Ruttan (2008) proposes that the type of heterogeneity at stake needs to be clearly specified (for instance sociocultural heterogeneity seems to have a more clearly negative effect than economic heterogeneity) and how success is measured has to be made clear – clarifying if success is measured in terms of collective action or in terms of level of collective goods provided (Ruttan 2008). A question remains whether to lump the effects of sociocultural and economic heterogeneity or study them separately. Naidu (2009) finds that in the presence of benefit heterogeneity, an increase in wealth heterogeneity reduces the extent of collective management. From the literature we find that physical characteristics of the resource and its associated usage are often given limited consideration (Araral Jr. 2009). For instance, in the case of a river, upstream-downstream (location) asymmetry and not economic heterogeneity may have a significant negative impact on collective action: the majority of poor water users located upstream may ignore a rich user located downstream, and the latter is less likely to influence collective action of the former. Similarly, powerful users (in terms of wealth) located upstream are also likely to ignore the downstream poor. Also noted by Bardhan and Dayton-Johnson (2002), an irrigation organisation crossing several village boundaries is less likely to rely on social sanctions and norms to enforce cooperative behaviour than that of a single village (users along a river have different incentive structures with respect to resource provision and appropriation). The impact of market opportunity also receives little attention. Araral Jr. (2009) finds that increasing market opportunity leads to increasing selfish behaviours among actors which lessen mutual dependencies, loosen up traditional social ties, and reduce the interlinkages for possible reprisals in the case of adverse behaviour.

Just like heterogeneity, the role of group size is also mixed. Although group theorists suggest that collective action is more difficult to achieve as group size increases, there is no consensus on how to establish the dividing line between small and large groups, or on the role of context in mediating the effects of group size (Poteete and Ostrom 2004, Araral Jr. 2009). An increase in group size is said to decrease opportunities for frequent interactions, lower reputation formation, and decrease expectation of future interaction thereby lowering the level of trust among users (Ostrom et al. 1999, Poteete and Ostrom 2004). Group size is said to affect the calculus and strategy of collective action even if trust is not a limiting factor (e.g. for some individuals the perception that an individual contribution does not make a difference increases with group size) (Poteete and Ostrom 2004). In addition, Poteete and Ostrom (2004) argue that as group size increases, threats of being punished in future become less effective as a method of encouraging cooperation. The logic is that an increase in group size increases the transaction costs of resource provisioning, thus raising the costs of initiating collective action. Hence, a large group is less likely to achieve collective action and if it did achieve collective action the level of resource provision will be much lower (Olson 1965). However, subsequent studies have shown that incorporating income effects leads to significantly different conclusions about the level of collective provision (Poteete and Ostrom 2004). Others argue that most collective goods are normal goods meaning that individuals who experience an increase in income decrease their expenditure on the goods by less than the amount of the increase in income.

In the following section, we present the case study context and the methodology used. This is followed by a section presenting the emergence and functioning of RCs in different parts of the Themi sub-catchment

7.4   RESEARCH METHODS AND CASE STUDY

7.4.1   Research methods

The objectives of this research were to describe and analyse the emergence and functioning of RCs in a sub-catchment. More particularly, we wished to understand when and why they emerged, how they evolved over time and their interface with government structures. To achieve these objectives, our research was premised on the assumption that network flows of water are dependent upon associations of humans, hydrological systems and the spatio-temporal construction of physical infrastructures. To identify the actors and their networks, we followed the water flow path downstream (Latour 1988, Law 1992, Murdoch 1998, Kortelainen 1999, Bolding 2004, Latour 2005) and in the process mapped the hydraulic infrastructures, the users and the institutional network behind them. To trace the emergence of RCs in the Themi sub-catchment, we first identified water users and their infrastructures and then mapped irrigation canals. Locally known as furrows, these divert water from the river and by gravity convey the water to the plots. Periodic maintenance is required to sustain the intake structures (often made of stone, tree logs and mud). Subsequently, furrow committees managing the furrows were identified and finally the RCs active in the sub-catchment. Following the result of spatial mapping and observation, we found that RCs are particularly active in Seliani and Ngarenaro tributaries and in the Lower Themi river.

Meetings were held, first with furrow committees and thereafter with the RCs of Seliani, Ngarenaro and Lower Themi. In Seliani river, meetings were held with four of the 12 furrow committees and one large-scale coffee estate. Two of the furrows were located in the upstream part of Seliani river, one in the midstream and one downstream. The second furrow upstream was recently constructed. For Ngarenaro river, meetings were held with all furrow committees. For the Lower Themi river, however, we only conducted meetings at the level of the RC, and not with the individual furrow committees.

Discussions with furrow committees were conducted as follows. The first part dealt with issues relating to initial investment in the furrow construction, current norms in use, how water is allocated to each furrow drawing from the same river and how this is transformed into water access for individual members of a furrow, understanding the efforts of the Pangani Basin Water Office to create catchment forums, and finally the acquisition of state-issued water rights. The second part of the discussion was on the RC: when it was formed, why it was formed and whose idea or initiative it was. Further questions included: how a furrow group can become a member of the RC, the specific role of this committee, especially with respect to water allocation, its spatial span of control, management structure and election of representatives and leaders. The discussion also focused on the link between the locally established RC systems with government-created water management structures and local government offices. Discussions with the RCs followed a similar format as described above for the second part of the furrow committees. In addition, the RCs were also asked to draw a sketch of the furrow systems under their command.

7.4.2   Case study sub-catchment

Biophysical context

Themi is a sub-catchment (Figure 7.1) of Kikuletwa catchment, which in turn forms part of the Pangani river basin, Tanzania. The sub-catchment covers parts of the districts of Arusha Rural, Simanjario and Arusha municipality, comprising 26 administrative wards with a total land area of about 363 km^2 (49% of the area of the 26 wards). Themi, Nduruma and Ngaremotoni rivers all originate from the slopes of Mount Meru (4500 m above mean sea level) and flow into the Shambarai swamp downstream (at about 800 m). Themi river is joined by four main tributaries, viz.

Naura Spring, Burka, Kijenge, and Ngarenaro rivers, the latter receiving additional waters from the Seliani and Burka rivers (see Figure 7.1).

Increasing water demand, water pollution and climate change make the area a potential hotspot of upstream-downstream water conflict. In the sub-catchment, users not previously aware of their mutual dependence on the limited water resource are increasingly being confronted with the need to share water with other users located distant from their area. The increased interdependency among water users has led to institutional innovation in the sub-catchment. Canal irrigation (furrow) groups have adapted their local institutions to the changing availability of the water resources. RCs have been created by farmers and made responsible for water allocation between groups utilising the same river.

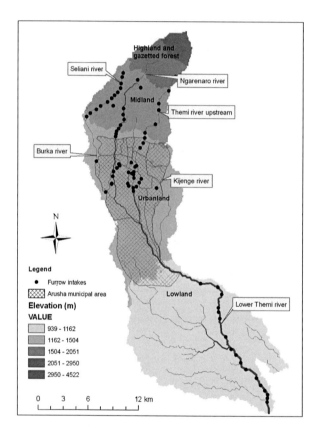

Figure 7.1: Themi sub-catchment river system and furrow diversion points.

The area experiences two rainy seasons per year, one starting in March and ending in May (locally called '*Masika*', also known as long rains) and the other starting around October and ending in December (locally called '*Vuli*', short rains). The average rainfall in the sub-catchment varies from about 1400 mm/yr in the highlands to about 500 mm/yr in the lowland. Inter-seasonal rainfall is relatively low (Figure 7.2).

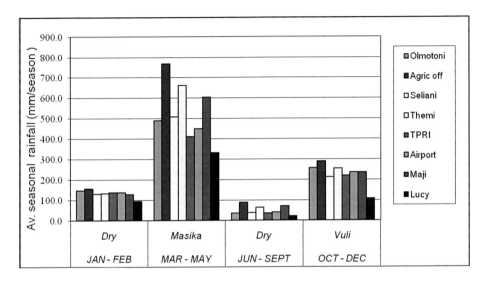

Figure 7.2. Seasonal rainfall as measured at different locations in the sub-catchment, 1927 – 1990; Olmotoni station is in the highlands while Lucy is in the lowlands (Source: Pangani Basin Water Office, Arusha).

Themi sub-catchment can be divided into four agro-ecological zones – highland/forest (over 1700 m a msl), midland (1500 – 1700 m), urban-land (1300 – 1500 m) and lowland (below 1300 m). The highlands (rainfall 1000-1400 mm/yr) comprise a gazetted forest reserve with some signs of human encroachment, timber logging and hunting.

In the mid-highlands (rainfall 800-1300 mm/yr) subsistence farming and stall livestock rearing are dominant. Although farmers practice rain-fed farming, canal irrigation is used for supplemental irrigation in the rainy season and full-scale irrigation during the dry seasons. Crops grown include maize, bean, banana, coffee, vegetables, tomato, and onion. In the mid-highlands, there is competition over water needed for the production of vegetables, tomato and onion that are increasingly in demand in the Arusha municipality.

Urban land comprises the built-up areas within the Arusha municipality (rainfall 600-800 mm/yr). Land use includes commercial activities and industries, although agriculture is practised on the outskirts of the municipality. Two coffee estates located in the outskirts of the municipality, and the small-scale farmers in this zone rely on rainfall, river water and municipal wastewater to grow crops. Water problems in this zone include pollution from municipal solid and liquid waste, as well as competition over water, especially between the surrounding villages and the large coffee estates. Furthermore, the increasing urban population puts pressure on the water supply, and the urban water authority (Arusha Urban Water Supply Authority - AUWSA) has to look for new sources both inside and outside the urban land area. This led to violent conflict between the water authority and the neighbouring villages in 2003 (Komakech et al. 2012e).

The lowlands receive limited rainfall (less than 600 mm/yr) and farmers here rely heavily on irrigated agriculture. Land use in the lowlands is dominated by subsistence farming and livestock rearing. Large sisal estates can also be found, but many of them have stopped operating, partly because of water scarcity. Whereas sisal is not normally irrigated, water is required for processing its products. In addition to water scarcity, pollution from the Arusha municipality presents a significant problem for lowland farmers.

Socio-economic context

The Themi sub-catchment is mainly inhabited by the Arusha people (a group of agro-pastoralists related to the Maasai), Meru (mainly agriculturalists related to the Chagga of Kilimanjaro) and Maasai. In 2002, the sub-catchment's total population was about 447,000 (Tanzania 2002b). The spatial geography is such that small-scale farmers upstream are confined between the forest belt, the urban centre and estates. Below the urban area and estates, smallholder farmers cultivate marginal lands with low rainfall and poor soil (see Spear 1997 for details on the origin of settlement in Meru). The location of water users close to the growing urban centre of Arusha - Tanzania's third largest city - has resulted in intense water resource use in the sub-catchment. Arusha city is at the core of a vital and relatively wealthy region with productive agriculture, mining activities and a large tourist industry, all of which are sources of intense water resources development and use (Carlsson 2003). In the sub-catchment, some rivers that used to be perennial have now become seasonal. There is a limited use of groundwater for irrigation, mainly by a few large commercial farmers. The Themi catchment is under the jurisdiction of Pangani Basin Water Board, a government- created basin organisation. The basin water board through its executive basin office (Pangani Basin Water Office - PBWO) is responsible for the allocation and management of water resources in the Pangani basin. So far, efforts by the Pangani Basin Water Office to enforce efficient water allocation remain weak[14] but institutional arrangements developed by farmers to secure supplies and share resources between upstream and downstream neighbours do exist and these arrangements mediate water access between the users.

7.5 EMERGENCE OF RIVER COMMITTEES

This section presents the findings from three RCs: Ngarenaro River Committee, which is comparatively well organised; Seliani Committee, where stiff competition has

---

[14] The basin water board currently grants fixed volumetric water rights to users based on assumed existence of average supply. The rights do not take into account seasonal variability.

reduced the committee's powers; and Lower Themi, where the committee experiences such fierce competition that the local government administrative structures are frequently called upon to settle conflicts. The RCs share a similar structure: they all have a chairman, a secretary, and water guards who are elected representatives of member irrigation canals and to a lesser extent commercial estates. This section describes the conditions under which the three committees emerged, their structure, roles and links, and how they attempt to allocate water and solve conflicts.

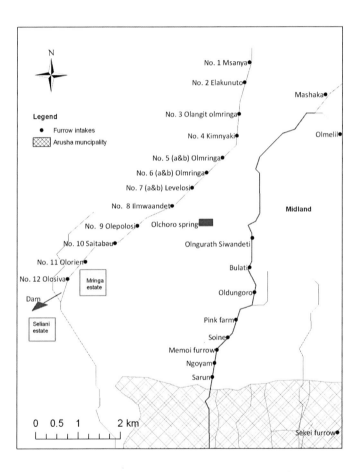

Figure 7.3 Seliani and Ngarenaro rivers showing furrow intakes (Source: field notes).

7.5.1   Ngarenaro river committee

The farthest upstream village using water from the Ngarenaro river is the Shiboro village. Altogether there are eleven furrows along the Ngarenaro river and its tributary upstream of Arusha (see Figure 7.3). The first furrows in Ngarenaro are said to have been constructed over two centuries ago for watering livestock, domestic water use and small- scale irrigation. The people of Arusha (*Waarusha*) reportedly contracted *Wachagga* from the Kilimanjaro region to construct the furrows. At the time, there was no competition over water and no institution responsible for water allocation between the existing furrows. Elders would inform the chiefs about the intended construction and mobilise labour.

Emergence of the river committee

Ngarenaro ('black water' in Maasai) River Committee (RC) was started around 1945 during a period of extreme drought. A group of elders from upstream and downstream divided the river into two zones, Oldungoro to Ilboru as upstream (referred to as *Oldungoro* by the farmers) and below Ilboru to the Ngarenaro confluence with the Burka river as downstream (referred to as *Burka* by the farmers), and each zone would use water for one week. Simple rules understood by the farmers as well as their leaders were put in place. The rules included fines and punishment for defaulters, and election of committee representatives.

To ensure a fair distribution, a measuring stick was introduced around 1960. The stick was introduced when it became difficult to continue with the weekly water allocation because of increased farming and livestock in the area. The idea of the measuring stick was partly triggered by events on the neighbouring Seliani river, where a farmer reportedly killed another over water. So the elders devised the system to avoid similar incidents on the Ngarenaro river. The measuring stick is kept by the RC chairman but the vice chairman also keeps a copy, which is used only for validation purposes.

The stick is made of bamboo measuring about 70 cm long, with four white markings (stages) used to measure water depth (
Figure 7.4). The first ring (from the bottom) corresponds to a water depth of about 11 cm, second one to 13 cm, third 15 cm and fourth 18 cm. Measurement is done at fixed points in each furrow, where the furrow width is about 70 cm (roughly the same as the length of the stick), and where there is laminar flow. Each of the graduations was set according to the population served, area to be irrigated and distance to the furrow command area (somehow taking into account transmission losses). The first three marks from top (water depth of 18, 15, and 13 cm) correspond to water allocation to three upstream furrows of Oldungoro, while the lowest mark (11 cm) is for downstream furrows of Burka. Olngurath furrow, having the biggest command area and traversing a longer distance to the irrigation command area than the other furrows, is allocated the uppermost mark. The second highest mark is for water allocation to the furrow with the second highest population, Bulati furrow. The third highest mark is allocated to the furrow with the third largest population, Oldungoro (Lesia) furrow, while the bottom-most mark corresponds to the water left for Burka's five downstream furrows (Pink farm, Soine, Memoi, Ngoyam and Sarun). The command areas of Burka's furrows are relatively small and are increasingly becoming built-up areas. These five furrows share the flow corresponding to a water depth of 11 cm on a rotational basis and the vice chairman of the RC is responsible for setting

the allocation schedule. Whenever there is sufficient flow in the river, water allocation proceeds according to the above procedure. But during periods of drought, water allocation is varied along the marking and depending on the severity of the situation, the markings are lowered for each furrow (e.g. furrow taking 18 cm is now given 15 cm, and the next given 13 cm, etc.); this is done until it is no longer justifiable to allocate water between the furrows. At this point, no one is allowed to irrigate anymore. The RC chairman, Mr. Nambua, said 'primarily we used the stick during water scarcity; no one is fined when water is enough. The period it is normally used is between January-March and August–October but effectively in the months of September–October, as this is the driest period in our area'.

Figure 7.4. The Ngarenaro RC chairman explains the marking of the measurement stick and demonstrates how it is used.

The committee's institutional structure

Each of the nine furrows is represented at river level by its chairman, vice chairman, secretary, and water distributors. The RC chairman, vice chairman and secretary are elected in a general assembly of all farmers (upstream and downstream). According to the present committee, election is in principle every five years or in case of a major problem, but this rarely happens. In the absence of a problem, the existing committee is maintained and new members join only when the furrow members change their leaders. The last general election was in 1995.

The RC is not registered with the Pangani Basin Water Office, but farmers believe it is nevertheless legitimate. No problems arise with the village governments since representatives of the village government are also members of the RC. Generally, the committee leaders are people holding several positions in the villages and wards. The current RC chairman, for example, is a clan leader and also a veterinary officer.

The roles of the RC include allocating water to member furrows, ensuring that all furrows receive their water turns, and calling for meetings when needed. The committee is also responsible for finding additional water sources during extreme shortage as well as identifying areas with high water losses, and negotiating new arrangements with upstream users who are not necessarily members of the RC. The committee is the only custodian of the measuring stick, and also represents the group in other fora including at the village government level and in other organisations, for instance those created by development NGOs.

Water allocation and enforcement

In Ngarenaro, water allocation between furrows still follows schedules developed in the 1960s. Individual farmers are prohibited from going to the intake unless it is for maintenance of the furrow intake which again must be done under the supervision of the furrow committee. Only furrow leaders are free to check and operate the intake structure. If a farmer has noted that flow into their furrow is significantly low, he must report to the furrow chairman who will discuss the matter with the RC chairman for a possible increase in flow at the intake. There is a minimum flow allocated to each furrow to cover domestic water needs and livestock. For each furrow, all water distributors (a water distributor or 'mgawamaji' in Swahili is a person responsible for day-to-day allocation of water to farmers within one furrow) are given a written water allocation schedule by the furrow chairman, which makes it easy to find defaulters on any particular day.

No serious conflicts are reported in Ngarenaro, but in case a furrow ignores the water schedule it is often fined a goat. The chairman stated: 'We do not need money; defaulters must bring a bull or goat and everyone enjoys. No fighting over water; conflicts must be brought to the leaders. Those who fight are severely punished; why fight over water when we have the leaders and the rules to follow?' Success of the committee is attributed to its consistency in applying the rules. A committee member said: 'If you give people freedom and later try to change it there will be trouble. Here, we frequently remind our members about efficient water use and about rules in place. If you are making laws do it together with everyone. In Ngarenaro, failure to follow set measurements is strictly punished (defaulters being fined a *dume* [bull]) and all furrow chairmen are responsible, and they must ensure that their intake is protected and members do not open the intake freely'. Pink Farm was reportedly fined a goat for pumping water; the committee thought the owner of the farm was taking more water than allocated. One other furrow, Olngurath, was fined a bull in 2008 for not following the schedules and the set measurement. However, the RC chairman also emphasises that humans do make mistakes; if this appears to be so, the case can either be dropped or the fine reduced to, say, a he-goat (*dafu*). Only Pink Farm is reported to have a state-issued water right but the committee said they plan to arrange for a water right at the river level and then distribute the right to the individual furrows. Current operational costs are met on demand (i.e. money is collected as and when needed). But some operational costs, such as communication costs are met by the individuals. "This is called 'commitment'," the chairman said.

The major challenge faced by the Ngarenaro River Committee is posed by the upstream village of Shiboro (Mashaka furrow), and Olmelil furrow, which are currently not members. Mashaka and Olmelil furrows say they are not aware of the

RC, and that their water is from a different branch of Ngarenaro not used by downstream farmers. However, they do remember the RC chairman coming to request them to allow night flow to downstream farmers. Shiboro village farmers think an RC is only useful when there is a water problem, but say that such a problem does not exist upstream at present. To overcome future upstream-downstream water conflict, Shiboro village is applying for a state-issued water right, which they think will control downstream interference.

7.5.2   Seliani river committee

The source of the Seliani river is located about 2000 m a msl in a gazetted forest area on Mount Meru. The river flows through the administrative wards of Ilkidinga, Kimnyaki and Kiranyi and terminates just before Burka spring on the outskirts of the Arusha municipality. The most upstream village is Shambasa, but this village mainly practices rain-fed agriculture. Two large coffee estates (Mringa estate and Burka estate, formerly Seliani, see Figure 7.2) also use Seliani water for irrigation, although since 2002 they have not received water. Upstream of Burka estate, 12 furrow intakes have been constructed along Seliani's course (Figure 7.3).

The people using Seliani river water include Maasai communities located upstream and a mixture of Maasai, Arusha, and Chagga downstream. This spatial geography of users has been pointed out by downstream farmers as a key obstacle to successful water allocation. Downstreamers are of the opinion that upstream Maasai communities are less likely to share water with the mix of downstream water users; using the Swahili word 'Mchanganyiko' (which refers to an area settled by people of different ethnicities). During a discussion with the upstream water users, a farmer said 'those people down there are not farmers; they are businessmen selling land to other people'.

Emergence of the river committee

Furrow number one (*Msanya*) is the oldest, constructed over 200 years ago, first for domestic water supply, fire-fighting and livestock watering and now being used for irrigation. By 1968, there were three furrows using water from the Seliani river. Five more furrows were constructed between 1968 and 1977. Later, four more furrows were constructed, the last one (furrow number two, *Elakunuto*) in 2003. According to farmers, water problems started in the 1950s when the owner of Seliani (now Burka) coffee estate (a Mr Isaac Blaumen) wanted to store water in a dam for his coffee plantation. The plan met stiff resistance from organised upstream farmers (mainly Maasai elders) who refuted Blaumen's claim of having a right to the exclusive use of Seliani water. The district commissioner was later involved and the right of the villages to water was formally recognised. An agreement on rotational water allocation was later reached, whereby upstream villages would use daytime flow and the night flow would be left for the Seliani estate.

It is this organised group of elders that evolved to form Seliani RC. In 1968, the farmers were led by someone named Laanoi, who remained the group leader for 30 years. The final decision to create an RC came around 1976. By then, more furrows had been constructed upstream and also the estates needed more water. There was another water conflict, and the three ward offices traversed by the river (Ilkidinga, Kimnyaki and Kiranyi wards), first assigned to manage the allocation between the estates and the farmers, could not manage the complex water allocation system. So the division secretary, who is at the next level of administrative authority (a division comprises several wards) called for another meeting to solve the downstream-upstream conflict. At the meeting, the parties agreed to create a committee that would bring all furrows and the estates together and be chaired by Mr Paulo Royand (chairman until 2001).

In 1976, the government warned that no more furrows should be constructed along the river. In addition, a restriction was put in place: to construct a new furrow the intending party would first have to consult the RC, a role previously performed by chiefs and clan leaders. It was also agreed that only the general assembly of the RC could decide on the construction of a new furrow. Despite the warning, more furrows were constructed. According to the current committee chairman, the RC has tried to control the construction of new furrows, but interference from the District authority has complicated matters. The chairman mentioned as an example that the District authority wrote a letter instructing the committee to allow the construction of furrow number two (*Elakunuto*) in 2002. Local politicians saw it as an opportunity to canvas votes in three wards, and furrow number two was built despite the expected negative impact on the system as a whole. Even then, upstream farmers consider their location near the water source to be synonymous with ownership of the river's water. Since they are favourably located upstream, they can ignore the RC: it was reported that the first three furrows (Msanya, Elakunuto and Olangit) did not attend committee meetings for two consecutive years. In a discussion, a downstream farmer remarked: 'We need upstream farmers but it is hard to bring them to negotiation terms with us downstream. (Although) we understand their advantaged position upstream, we will guard our water turn whenever possible.'

Water allocation and enforcement

Water is allocated on a rotational basis between the furrows. Initially, the villages would share daytime flow, while night flow was for the Seliani estate. Before 1970, allocation to the villages (only three furrows then) was from 6 am to 6 pm: so each of the three furrows could get water for three consecutive days (sufficient for the existing water use). Failure to comply was a fine of a bull. There was strong leadership by the chairman, Paulo Royand, who could enforce the rules. In addition, Seliani estates would meet the cost of furrow rehabilitation and meetings. To reduce water theft, the Seliani estate constructed intake gates with locks on all the furrows upstream and employed a water guard to open and close them. But this never worked, as farmers bribed the guard and water could be used outside the allocated hours. Eventually, the farmers kicked out Seliani and Mringa estates from the RC, citing increased water demand (Box 7.1).

According to the RC chairman, by the year 2000 it would have taken 36 days to complete the water allocation cycle. But since the two estates were kicked out of the RC and allocation to the furrows was restructured to day and night it now takes 21 days to complete an irrigation cycle. Current water allocation schedules are being prepared by the RC secretary starting in January each year. In the schedule, upstream and downstream furrows are paired and given one day allocations, which they share equitably (12 hours for each). The paired furrows take turns irrigating (upstream furrow starts, e.g. during their turn, furrow number 6 starts, irrigates for 12 hours and leaves the water for furrow number 12). Schools within the area are allocated water one day per week (basically Saturday 6 pm to Sunday 6pm).

Box 7.1: Statement of Burka Estate Personnel Officer.

"Sharing water with smallholder farmers is challenging. Commercial farmers are often seen by small farmers as different and foreign. In meetings, farmers often shift to their local language. How can a commercial farmer have meaningful communication with small upstream users speaking a different language? We used to be members of the Seliani river committee together with Mringa estates but upstream farmers cut us off in 2002. Luckily the spring source feeding the Burka river is located within our farm. We have protected the spring source, invested in groundwater and have been innovating with our irrigation technology. Currently, we are using a variant of drip irrigation, PIDO. It is a simple technology that can easily be moved from place to place".
Source: field notes

Enforcement of the water allocation schedule remains a problem. Water theft upstream, particularly by the first, second and third upstream furrows, as alleged by many downstream irrigators, makes the RC ineffective to downstream users. According to the general rule, violation of the schedule would attract a fine of Tshs70,000 (approximately US$45) but enforcing this rule has proved difficult. There is also an ethnic/tribal dimension: upstream farmers, being mainly Maasai, often speak as one and tend to band together against the mixed ethnicity downstream. Another critical factor is land use change upstream. Upstream Maasai used to be cattle keepers with minimal water demands. But as they have taken to growing crops their water demand has increased significantly. The impact of the RC ineffectiveness is felt downstream, by furrow numbers seven through twelve. Here, irrigators sleep out guarding their water turn. A farmer from furrow number ten stated:

> *"When it is our water turn, we go upstream at 5pm to close furrow intakes and remain guarding the intake until 7am in the morning. Unfortunately, due to transmission losses in the dry riverbed, 7am is also the time water normally reaches our farms downstream and yet by 4pm upstream farmers will have opened their intakes. In fact, during our water turn, we only get water for about 10 hours instead of the official 24 hours allocated."*

Seliani RC also does not have a formal link with Pangani Basin Water Office (PBWO). Some of the users (e.g. Seliani coffee estate) have acquired water rights from PBWO but the rights are not used in the allocation of water or recognised by other water users.

7.5.3   Lower Themi river committee

Themi river also rises on the slopes of Mount Meru above the forest belt. Its source is called *Emaoi*. The river flows through the villages of Oldonyo Savuk, Kivulul, Moivo, Sekei, and Arusha municipality before being joined by Kijenge and Burka tributaries. Through its course, Themi connects water users from the Arumeru district, Arusha municipality and the Simanjiro district.

Upstream of Arusha municipality, water is abstracted by three furrow irrigation systems (one in Kivulul village and two in Moivo village) and by Arusha Urban Water Authority. There is no RC in this upstream section of the river. In the village of Kivulul water allocation is managed by the village government through elected representatives from sub-villages served by the furrow. The village executive officer prepares a water turn as a written paper (*'Kibali'*) and gives this to the sub-village representative, who issues the water at a fee of Tshs100 per turn (approximately US$0.07). The two furrows of Moivo village do have committees which follow the traditional Maasai age system (called *'Jando'*). At present, all three furrows divert water at the same time. The villagers of the Moivo village believe that the system functions without a central authority because the traditional rule that the river may not run dry since the water is also needed by aquatic animals is still respected; hence none of the furrows divert all river water.

A similar arrangement is found midstream within Arusha municipality. Some of the furrows in this area rely mostly on waste-water effluent and spring sources. Farmers in the midstream are affected more by water pollution from industries such as Tanzania Breweries (a beer factory that uses caustic soda for bottle cleaning).

More furrows – 40 km downstream of the municipality of Arusha – serving the wards of Nduruma (Kichangani village), Bwawani, and Oljoro (Simanjiro district) have been constructed and an RC has been created to coordinate water allocation (Figure 7.5). Water is mainly used for irrigation, as well as for livestock watering. The area also has the largest livestock population in the whole of Arumeru district and water users from Simanjiro district are also livestock keepers. There are two sisal estates each using Fili and Lucy furrows.

As the Themi river passes through Arusha municipality it becomes heavily polluted with industrial and municipal waste, and pesticides used by upstream farmers. Nevertheless, its water is used for drinking, cooking and washing by downstream communities including secondary schools. All the wards except Nduruma use Themi water for domestic water supply.

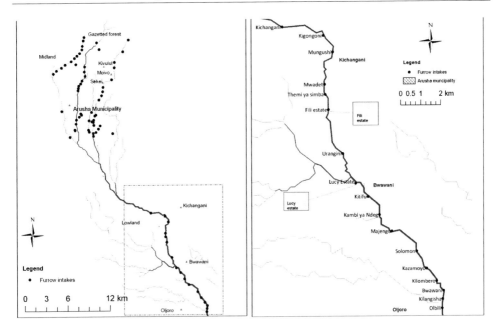

Figure 7.5: The left panel shows Lower Themi position (indicated by the dotted rectangle) relative to upstream users like Arusha city and Kivulul village. The right panel shows Lower Themi furrows (Source: Field notes).

Emergence of Lower Themi RC

Prior to 1992, there were not many water users in Lower Themi. Due to in-migration and natural population growth, more land was put under cultivation, furrows were extended and new furrows were added, thereby increasing water shortages. The idea of an RC started around 1992 when the river flow was no longer sufficient for all furrows to divert at the same time. There were conflicts over water and the division secretary, a retired military officer, together with the village elders, initiated the RC system to overcome the problem. The first agreement was verbal, with no written rules. But in 1999, another meeting was held between the local government and the RC, during which byelaws were drafted, forming the RC constitution, copies of which were distributed to all the ward offices. According to the committee secretary, people are often motivated to work together when there is not enough water and this was the driving force for creating the RC. The estates downstream were given orders to join the committee and because water is required for sisal processing, they were motivated to become members. According to the farmers, the RC is essential in resolving water conflicts during water scarcity. The committee produces an allocation schedule for all 17 furrows in Lower Themi (i.e. from Kichangani to Olbili furrow, indicated in Figure 7.5, right panel).

Water allocation and enforcement

During the rainy seasons, water is not formally allocated. According to the chairman, during *masika* (long rains), sufficient water is available to satisfy all needs, so all furrows can get water each day and still some balance remains in the river. During the dry season, allocation is based on turn-taking. Depending on water availability, upstream furrows use water for 2days and downstream furrows for the next 2 days. Rotational allocation is reported to start in July and end on the 30 September every year. After September until the onset of the *vuli* (short rains) season, no one is allowed to divert water for irrigation purposes. The flow is left in the river to be used by livestock. Whenever there is an increase in river flow (e.g. from early rains upstream), the committee revises the allocation schedule. Depending on climatic conditions, rotational allocation may start and/or stop early or late. In August 2009, while mapping the furrows together with the RC Secretary, an elderly Maasai livestock keeper complained that the ban on irrigation use needed to be brought forward to 31 August. He said their livestock had been without water for 2 days and threatened to send young Maasai boys upstream to destroy all furrow intakes if they did not get water. These are not mere threats; in 2008, Maasai youth reportedly destroyed all furrow intakes, and also destroyed crops along Kichangani and Kigongoni furrows.

The allocation schedule is enforced by water guards who routinely patrol the river to prevent violation. Each water guard is paid Tshs4000 (approximately US$2.50) per day for the work and two or three water guards are deployed every day. Water guards are reportedly selected by the villages and elected in a general assembly (each village nominates two names, one of whom is elected). Presently, all the water guards are from downstream, as upstream villages are reluctant to pay for the water guards.

Some of the furrows have a government granted water right (e.g. Kichangani, Mungushi (expired), Kigongoni, Themi ya shimba, and Fili and Lucy estates). The other furrows are in the process of acquiring water rights. Water is allocated according to local norms and in case of a problem, the committee assumes responsibility. Holders of government issued water rights are not given priority or favoured and the fixed discharges defined in these permits are ignored.

Along the Lower Themi river, there are frequent conflicts over water. According to the committee, there were conflicts in 2005, 2008, and 2009. In all of them, the police got involved. Problems arose from lack of water downstream and people often destroyed intakes and sometimes crops or even beat up farmers. The allocation system is reported successful only when there is sufficient water in the river. But during extreme scarcity, some users, especially upstream furrows, divert water without respecting the water schedule set by the committee. The RC secretary stated that it is often the upstream furrows of Kigongoni, Mungushi and Kichangani that cause problems during water shortages. The Kichangani and Mungishi furrow intakes are problematic, as the whole river needs to be blocked before water can enter these furrows, thereby affecting downstream users. The intakes also require several man-days to reconstruct, so whenever downstream farmers go to clear the obstructions created by these intakes, conflicts arise. During interviews, we were told that the committee had taken Kichangani furrow to the government court because of violating the schedule. We also observed that one of the large sisal estates used its power to influence the RC and a nearby local police post. They often pay the water guards to patrol the river and buy fuel for the motorcycle used by the police to arrest upstream violators.

A traditional fine system is supposed to be in place; a bull or its equivalent of Tshs 100,000 (approximately US$65) is levied when a furrow closes the river completely thereby causing water shortages downstream. The RC has tried to enforce this rule, but no one pays. The committee observed that it was not easy to get money from people. According to the committee secretary, this is because the upstream furrows have been in existence for a long time; moreover, some farmers have acquired water rights from the Pangani Basin Water Office and used to pay annual water use fees to the central government (currently, no furrow pays the yearly water fees); finally, some of the upstream furrows are located in a ward of another district which makes it even more difficult for the committee to fine them. 'So we have to accommodate upstream disturbance somehow. Sometimes we destroy the intakes but also we do understand that it takes so much time and labour to reconstruct'. Downstream furrows are not lined and use traditional intakes made of stones, mud and logs. Equipping the furrow intakes with gates is considered by the Lower Themi RC as a solution as these are believed to reduce leakage and improve effective allocation to the furrows.

During mapping, we noticed that some of the furrows, which previously had intakes with lockable gates installed had been destroyed. When asked why they could not maintain the weirs, the RC secretary claimed the intakes, constructed between 1992 and 1994 when the government introduced the water rights system, were destroyed by the farmers because they did not agree with the government water rights and ownership system. He also said, however, that farmers now understand the system well, and that a recent change from the water rights system to a water use permit system is considered acceptable. The new Tanzania Water Act of 2009 abolished the system of granting water rights and instead introduced water use permits. The permits grant access to beneficial use of water and does not confer full ownership over the water (Tanzania 2009).

Operational costs of the RC are met by water user's contributions and fines. These contributions are derived as follows: every member furrow contributes Tshs10,000 per year (approximately US$6.50). The furrows get the money from their individual member farmer's contribution (normally between Tshs1000-2000 about US$0.65-1.30). But most money is collected on the spot whenever there is a need. Each furrow chairman has to contribute Tshs 4000 (approximately US$2.50) per day for paying water guards. Since the committee is not formally incorporated as an association, it does not have a bank account; any balance is kept by the treasurer.

Currently, the Lower Themi RC does not communicate with users in the upper part of Themi sub-catchment. In 2008, the committee members visited the Pangani Basin

Water Office (PBWO) sub-office in Arusha to find out how they can reach upstream users. PBWO advised that the committee would be called for a meeting with the other users in early 2009. By the time of our interview, the PBWO had just started consultation processes over the creation of the Kikuletwa Catchment Forum of which Themi sub-catchment forms part (discussed in Chapter 4). Catchment fora are envisaged to provide an arena for users to dialogue on water allocation and management issues at an intermediate level (between furrow and basin).

7.6   Discussion: Emergence and functioning of river

Committees

This section discusses the drivers for the river committee formation, and assesses the emergence of these institutional arrangements using the concept of institutional bricolage. It then analyses the RCs endurance over time using Ostrom's eight design principles. In addition, we examine the impact of heterogeneity and group size on the functioning of the RC system.

From the preceding accounts, four drivers can be identified as the main triggers for the RC formation in the Themi sub-catchment: (a) increased frequency of low flows in the area, which has increased competition and sometimes violent conflict over water; (b) natural population growth, which put more pressure on land and water resources in the catchment (see Mbonile 2005); (c) the availability of markets for agricultural produce– Arusha, being a fast growing city with a good road network to Dar es Salaam and neighbouring Kenya, provides market opportunities to nearby farmers and this in turn leads to agricultural intensification and competition over water resources; and (d) colonial and, later, the independent Tanzanian government policies; for instance, the Tanzanian government 'ujamaa' in the 1970s settled pastoralist Maasai, especially along the Seliani river and intensified agricultural water use leading to fierce competition with the downstream coffee estate. These four factors contributed to rendering existing local water sharing arrangements ineffective. Thus, institutional innovation became a necessity and the RC emerged to solve coordination challenges that the ward offices could not.

The concept of institutional bricolage contributes to explaining the creation process of RCs in the Themi sub-catchment. The RCs benefited from the already existing arrangements such as the principle of good neighbourliness, the rationale of local water allocation, and traditional conflict resolution mechanisms. This underscores the argument that new institutions often benefit from the legitimacy of past arrangements (Cleaver 2002). The aspect of the multiple identities of the bricoleurs is well illustrated, as the creation process included, among others, local politicians and the district commissioners (DCs). The DC reports directly to the president's office and his/her responsibilities include maintaining security and peace in the community. Although it was always the downstream users that initiated cooperative arrangements, villages, wards and division offices played significant roles in the creation of the committees as well, e.g. through the efforts of division secretaries, which underscores the importance of non-users in local water conflict resolution. The farmers themselves have multiple identities and roles. Adaptation of the traditional age-group system (for example, the Maasai '*jando*' system, where certain age groups are responsible for resource management) to manage water allocation between villages is illustrative of the multipurpose nature of local institutions and of cultural borrowing as argued by Cleaver (2002). Although bricolage provides for understanding how such a change occurred, it does not answer why it was necessary to adapt existing arrangements at some point in time. Nor does it explain how the institutions function. To understand how the RCs endure over time we assess the three case studies using Ostrom's eight principles of long-enduring self-governing local institutions (Table 7.2).

Unlike Ostrom's claim that well-defined boundaries are needed, in Themi the boundary definitions used are context-dependent, ambiguous and fluid. In all three cases, boundaries follow the extent of the downstream users' claim to water at the time, as the three RCs comprise currently only members from a maximum of three administrative wards even where the river is also being used by other wards (Ngarenaro and Themi being good examples). Membership is not entirely closed to outsiders. Non-members can seek permission and, if allowed, pay entrance fees - a mechanism that lowers the cost of controlling the boundary. Though not necessarily the intention at the time, by leaving the boundaries open to negotiation future demands can be met in the face of increasing resource variability (e.g. in Ngarenaro one of the RC's roles is to negotiate with upstream users who are currently non-members in cases of extreme shortages). The boundary principle is frequently violated by upstream users, as illustrated by the case of Seliani, where upstreamers and their local politicians invested in more furrows without the consent of the RC. It is not immediately clear whether the ambiguous boundaries of all three RCs studied should be considered a weakness. However, given that one RC performs quite well (i.e. Ngarenaro) demonstrates that having well-defined boundaries is not a universal condition.

The second principle of congruence between appropriation and provision rules and local conditions applies, but only to a certain extent. The RCs have developed varied levels of water allocation. The Ngarenaro committee, for example, invested in proportional allocation using a marked measurement stick and clear rules on its use as well as punishment for defaulters. The other RCs (Seliani and Lower Themi) have continued with rotational allocation but as the number of furrows increases, the schedule becomes increasingly complex leading to an increase in water theft, which triggers the involvement of the national police. Although the government's system of water rights is well known to the actors, it is not being used to guide water allocation. Instead, farmers seek to acquire water rights only to strengthen their claim of ownership, but actual water allocation does not follow the government-allocated right. Given the rules developed by the RCs, upstream users are expected to forego immediate benefits without receiving (direct) compensation from downstream users. This, however, is not easily enforced, which culminates in frequent water theft and rule violation by upstream users as seen in the case of Seliani and Lower Themi RCs. Thus, the second principle may partially explain Ngarenaro's success.

Table 7.2: Applying Ostrom's design principles to the three river committees.

Design principles	Seliani RC	Ngarenaro RC	Lower Themi RC
1. Clearly defined boundaries (of resources and users)	*Resources:* All administrative wards crossed by the river involved. *Users:* Initially, all users were involved but two large downstream users were removed by the upstream users, citing increased water demands upstream.	*Resources:* Not all wards crossed by the river are involved. Boundary remains flexible, and is often enlarged when water gets scarce. *Users:* Not all furrows are members. Downstream furrows within Arusha and the most upstream village (Shiboro) do not participate.	*Resources:* Upstream and midstream section not involved in the downstream RC. *Users:* Only users from the lower section of the river are involved in the RC. Upstreamers have developed their own arrangements.
2. Congruence between appropriation and provision rules and local conditions	Water allocation is by turn-taking but individual farmers' guard their water turns.	Uses a robust proportional system to allocate water – clearly marked stick used and violators fined.	Allocation is by turn-taking but water guards paid by downstreamers monitor allocation. Police often involved in settling conflicts.
3. Collective choice arrangements	Large estates excluded by upstream farmers. Rules not always enforced. Frequent theft. Upstream reluctant to participates in modifying rules.	Not all affected are involved, but existing groups strictly follow the rules-in-use. Cheating is punished with a fine. First time offenders often pardoned.	Frequent rule violation occurs upstream, conflicts emerge and police frequently involved.
4. Monitoring	RC is responsible for monitoring but ineffective. Initially, the downstream estate employed a guard. Farmers do self-monitoring by going upstream.	Chairman of the RC and his team are responsible. But all furrow members monitor water use, who report flow reduction to the chairman.	Water guards are employed to monitor water use along the river. The guards are also farmers, mainly from downstream, and paid by other users.
5. Graduated sanctions	Defined but proved difficult to implement. Violation of schedule would face a fine.	Clear sanctions in place and being followed. Minor violator warned or fined a goat, major ones fined a bull.	Clear sanctions defined but not in use; instead government police is drawn in.
6. Conflict-resolution mechanisms	RC is responsible for conflict management but is sometimes ineffective; local politicians also involved.	RC is responsible for conflict management. If unsuccessful, other forms of traditional conflict-resolution methods are used. If this also fails, the case is taken to local government offices but this rarely occurs.	RC is responsible for conflict management. Clear procedures are defined but not strictly followed. Some users go directly to government courts, others resort to violent conflict, and still others use their resources to pay water guards and/or local police and access water outside their turns.

Design principles	Seliani RC	Ngarenaro RC	Lower Themi RC
7. Minimal recognition of rights to organise	The RC's right to organise is strongly affected by local politics. Ward councillors and executive officers interfere with RC operations. RC is not formally recognised by the basin water board.	Traditional leaders play a strong role in RC operation. Local politics not an issue. RC is not formally recognised by the basin water board but some of the leaders are government employees.	The RC right to organise is strongly affected by local politics. Ward councillors and executive officers interfere with the RC operations. RC is not formally recognised by the basin water board.
8. Nested enterprise (hierarchical or interrelated organisation levels)	Nesting is not uniform. RCs comprise elected furrow representatives. Allocation systems vary between furrows. Higher local-administrative levels are involved in an ad hoc fashion during conflict.	Not all furrows involved. Mixed within traditional structures. Leaders are furrow representatives. Not linked to higher water- management levels. Local administration involved in an ad hoc fashion.	Leaders are furrow representatives. Not linked directly to any administrative or higher water management level. RC only responsible at the river level and not involved in individual furrow management.

The third principle (collective choice arrangements) is not strictly adhered to, as not all the river users are members of the RCs, and those who are currently members, particularly upstream, may still object to rule changes by absconding from meetings (as illustrated by the case of Seliani where upstream users refused to ratify a new constitution).

Monitoring (principle four) within these systems is an important activity in all RCs but not systematically organised. Farmers do self-monitoring but in some cases guards are hired and paid exclusively by downstream users. This is illustrated by the cases of Seliani and Lower Themi where farmers have to invest extra resources such as money, time in guarding water or even force (e.g. destruction of upstream furrow gates by downstream livestock keepers) to secure allocation.

Although mechanisms for graduated sanctions (principle five) are in place, frequent violations occur even in the most stable RC (in this case Ngarenaro). Other factors such as the presence of markets seem more important than trusted relationships - for instance, the demand for agricultural produce (e.g. vegetables) in Arusha municipality promotes a different kind of water rationality (cf. Alam, 1998), in that upstream users now value water more as a commodity than as a common good, loosening underlying principles such as good neighbourliness.

We found that with respect to conflict resolution mechanisms (principle six) only one of the three RCs largely succeeds in managing conflicts itself. The conflict management capability of the other RCs is limited and conflict management is often quite a messy process, whereby sometimes individuals take cases directly to the police or the courts, and in which local politicians frequently interfere.

On the need for minimum recognition of the users' right to organise (principle seven), it may be concluded that such recognition occurs differently at the different administrative levels of government (village, ward, division, district, region and state), that the RCs as such have not been formally recognised administratively, but that the National Water Act of 2002 does recognise the water users' right to organise themselves at the river level.

Nesting (principle eight), finally, is poor as not all catchment users are considered and there is no formal link with both the larger basin organisation and the national policy level. Since only some of the users are involved, an RC can best be described as a structure for bridging relations between competing wards rather than as a nesting structure of all water users of a particular river.

The foregoing discussion can explain the performance of the most successful RC (Ngarenaro) by referring to five of Ostrom's eight design principles: (1) congruence between appropriation and provision rules and local conditions; (2) monitoring; (3) graduated sanctions; (4) conflict resolution mechanisms; and (5) minimal recognition of rights to organise. Apparently, not all eight design principles are required for a water institution to be effective. However, it remains unclear why even the two less-successful RCs, which lack clear collective choice arrangements, and effective mechanisms for sanctions and conflict resolution, have endured over time and have not collapsed.

The difference in the performance of the three RCs cannot be explained by Ostrom's eight design principles alone. We also need to consider factors related to heterogeneity as reported in the CPR literature. Table 7.3 profiles the three cases against factors of heterogeneity and group size. The following points related to the impacts of heterogeneity and group size on collective action can be drawn from the cases.

We find that location (hydraulic) asymmetry, and not economic heterogeneity (wealth), impacts negatively on the ability of river users to maintain well-functioning collective action institutions. All the upstream users in the three cases interviewed believe that their advantageous location is synonymous with ownership of the water and a licence to use more. It is even more complicated to enforce local water allocation rules with increasing external markets that foster individualism among actors thereby lessening mutual dependencies, loosening traditional social ties, and reducing the interlinkages for possible reprisals in the case of adverse behaviour (Araral Jr. 2009). In Seliani river, upstream ownership claims have become much stronger with the availability of attractive and reliable markets for agricultural produce, and upstream users say they know the monetary value of water. It can be concluded that a rich user located downstream is less likely to influence collective action when the majority of the users are located upstream, even if they are comparatively poor (Seliani RC is a good example): here the downstream Burka coffee estate, despite its wealth of resources, has not been able to skew water allocation to its advantage. The estate constructed lockable furrow gates upstream and employs more than 400 farm workers from upstream villages but this did not pave the way for cooperation from the less-wealthy small-scale farmers upstream.

Instead, estates were seen as foreign and different. In Lower Themi the downstream sisal estate often provides fuel to government police and bribes committee leaders to get more water. In addition, the heterogeneity of the users (e.g. in terms of ethnicity) seems to play a critical role in the instability of water-sharing arrangements. Seliani users cite the coalition of upstream Maasai/Waarusha users as the cause of the RC's weakness, while the Ngarenaro committee is said to be effective because of its relatively homogeneous Waarusha user community.

Table 7.3: Comparison of the cases in terms of heterogeneity and group size.

Factors	Seliani	Ngarenaro	Themi
Group size - number of users sharing a tributary	12 furrows	8 furrows	40 furrows in total, one urban water supply intake and about 17 furrows in the lower Themi river
Heterogeneity of wealth – differences between the users in terms of size of water use (commercial vs. subsistence)	Mostly small-scale subsistence farmers. Two large coffee estates	Mostly small-scale subsistence farmers. One estate that is beginning to fail	Majority are small-scale subsistence farmers; Arusha water supply, two sisal estates but one is out of production
Heterogeneity of type of water use (irrigation, livestock, domestic)	Irrigation, domestic, livestock	Irrigation, domestic, livestock	Irrigation, livestock, domestic, urban water supply
Membership of communities, administrative units and different ethnic groups	Maasai, Waarusha, Chagga	Mainly Waarusha	Mixed communities, formerly migrant workers to sisal estates
The spatial distance between the users (most upstream and downstream)	8.0 km	8.3 km (14 km between the most extreme users of the Ngarenaro river)	15.0 km (45 km with most extreme user of the Themi river)
The institutional distance between the users	Three wards, all within one district	Three wards, all within one district	Three wards from three districts

In line with the above, we conclude that an RC crossing several villages is less likely to rely only on social sanctions and norms to enforce cooperative behaviour than that used by a single village. This is illustrated by the Lower Themi committee with users from three wards, each from another district, which relies more on the government courts than on local conflict-resolution mechanisms to solve water conflicts. Bardhan and Dayton-Johnson (2002) reported similar findings for irrigation organisations in Nepal, southern India and central Mexico.

Finally, although the RCs are active in the same sub-catchment, they operate independently of one another, and more surprisingly, they do not presently communicate with one another. Hence, there is no awareness of how the other committees operate, foreclosing possibilities for mutual learning. It may be hypothesised that the larger the spatial extent between upstream and downstream users, the more difficult it is for such institutional arrangements to emerge from bottom-up. The RCs work more closely with the village and ward offices and are not formally linked to the official basin administrative structures. It is only recently that the Pangani Basin Water Office has tried to establish nested water management structures (e.g. water user associations created under sub-catchment fora/committees) but again, the new structures do not explicitly aim to build on or start from the RCs; they often duplicate the roles and functions of these locally developed arrangements.

7.7 Conclusions

This chapter set out to describe and analyse the emergence and functioning of RCs in a sub-catchment of the Pangani river basin. In particular, it discussed when and why they emerged, how they evolved over time and their interface with government structures. RCs in the Themi sub-catchment emerged in response to competition and conflicts over limited water induced by, among others, increased frequency of low flows, natural population growth, markets for agricultural produce, and government policies. It is notable that the idea to create an overall and self-governing body responsible for water allocation along a particular watercourse was partly the work of parties who were not direct resource users (e.g. the District Commissioner, and Regional Commissioner) but who directly had interests in the peaceful coexistence of the user community. This notwithstanding, the RCs were created on the principle of already existing arrangements and thus benefited from the legitimacy of past arrangements (cf. Cleaver 2002). Currently, the committees operate in a pragmatic manner, depending on situation-specific climatic and hydrological conditions. They are active during dry seasons and in times of extreme water shortages, but in the event of (unexpected) rainfall even planned meetings may be abandoned.

Ostrom's eight design principles provided useful entry points in studying the functioning of the RCs. The best performing RC in the Themi sub-catchment largely complied with five of the eight design principles (Ostrom 2002), which indicates that not all these principles are necessary for a water institution to be effective and to endure over time. The other two RCs, which performed less effectively, only adhered to three of the eight principles. All three RCs, however, leave the resource boundary open to negotiation. Boundaries are fluid and change over time and are thus not clearly defined. By not entirely closing the resource boundary, the users lower the transaction cost of controlling the boundaries and also allow future demands to be met in the face of increasing resource variability. In addition, the success of the boundary definition is dependent on the capacity of the committee to motivate upstream users (Seliani and Lower Themi cases illustrated how upstream users frequently violated the rules and constructed more furrows or used water outside their turns). All three RCs also do not fully comply with the principle that all affected must take part in rule creation and modification. Moreover, not all the water users are currently members of the three RCs. Also, members may consciously abscond from meetings to avoid ratifying binding agreements. Monitoring is at best ad hoc as downstream farmers have to invest extra resources to secure allocation. In all three cases, finally, the "nesting" of lower-level institutional arrangements within higher-level ones was inconsistent, but this did not necessarily hamper their functioning.

We could not explain the difference in the performance of the three RCs by referring to Ostrom's eight design principles alone. We also needed to consider factors related to heterogeneity, which were shown to be putting the RCs' operations under increasing stress. The functioning of RCs is strongly influenced by: (1) the number of users sharing a tributary; (2) differences between the users in terms of type of water use (irrigation, livestock, domestic) and size of water use (commercial vs. subsistence); (3) sociocultural differences between the users, e.g. different ethnic groups; (4) location (hydraulic) advantage and spatial distance between the users; (5) the crossing of administrative boundaries; and (6) the presence of markets for (high-value) agricultural products.

The entrepreneurial farming opportunities in Arusha municipality promote a new kind of water rationality. Whereas in former times water users seem to have defined their self-interest in terms of broader social, spatio-temporal interdependencies (Van der Zaag 2007, after Alam, 1998), this is now changing. Upstream users now view water as a source of private wealth, rather than a resource that requires collective action to generate a stream of benefits.

Plastic valley: view of flower estate in the midland of Nduruma

Chapter 8

THE ROLE OF STATUTORY AND LOCAL RULES IN

ALLOCATING WATER BETWEEN LARGE AND SMALL-

SCALE IRRIGATORS IN AN AFRICAN RIVER

CATCHMENT[15]

8.1 ABSTRACT

This chapter presents a case study of large and small-scale irrigators negotiating for access to water from Nduruma River in the Pangani River Basin, Tanzania. The chapter shows that despite the existence of a formal statutory water permit system, all users need to conform to the existing local rules in order to secure access to water. The spatial geography of Nduruma is such that smallholder farmers are located upstream and downstream, while large large-scale irrigators are in the midstream part of the sub-catchment. There is not enough water in the river to satisfy all demands. The majority of the smallholder farmers currently access water under local arrangements, but large-scale irrigators have obtained state-issued water use permits.

---

[15] Based on Komakech et al. 2012. *Water SA* 38(1), 115-126

To access water the estates adopt a variety of strategies: (1) they try to claim water access by adhering to state water law; (2) they engage with the downstream smallholder farmers and negotiate rotational allocation; and/ or (3) they band with downstream farmers to secure more water from upstream farmers. Estates that were successful in securing their water access were those that engaged with the local system and negotiated fair rotational water sharing arrangement. By adopting this strategy, the estates do not only avoid conflict with the poor downstream farmers but also gain social reputation, increasing chances of cooperative behaviours from the farmers towards their hydraulic infrastructure investments. Cooperative behaviours by the estates may also be due to their dependence on local labour. We further find diverging perspectives on the implementation of the state water use permits – not only between the local and state forms of water governance, but also between the differing administrative levels of government. The local governments are more likely to spend their limited resources on "keeping the peace" rather than on enforcing the water law. At the larger catchment scale however, the anonymity between users makes it more difficult to initiate and maintain cooperative arrangements.

8.2 INTRODUCTION

Increased water scarcity leads to competition between water users, large and small, up and downstream. Conflict may arise because upstream users abstract most of the water and leave their downstream neighbours with scarcity. To solve water allocation conflicts, many governments attempt to formalise the water right system - users are granted rights to use a certain amount of water, at a particular location and duration. However, the formalisation of water rights may also provide opportunities for wealthier, more powerful, and better connected users to manipulate registration to serve their own interests (Bruns 2007). In addition, since sources of water rights are multiple and often conflicting, formalisation may lead to struggles over whose water right is legitimate. Smallholder farmers may base their water claims on customary rights and their historical investments in water infrastructures, while new users (e.g. large-scale irrigators and cities) use state-issued water rights to gain control of water sources.

This chapter presents the struggles for water access and control in Nduruma River, upper Pangani River Basin, Tanzania. The present water rights system in Tanzania builds on water law established by the colonial authorities in the early 20th century – a law specifically designed to limit use among native inhabitants while at the same time securing access to water for European settlers (Lein and Tagseth 2009). In addition to the colonial-induced water access asymmetry, more recent increased water demand has led to fierce competition in the Nduruma sub-catchment. This increasing water demand is partly caused by the revitalization of coffee estates by both local and international private capital. Several of these estates are relics of the German and British colonizers (cf. Spear 1997). Most of the coffee estates have been converted into large flower farms by a new group of white farmers, making the present social-geography a mirror image of the colonial past. The area which was once called the 'iron ring of alienated land' (Spear 1997), is called today the 'plastic valley' because of its numerous greenhouses. Old estates were also the centre of the protracted struggles over water and land resources between the local people (Meru and Arusha), white settlers, and the colonial administration (Spear 1997). Just as in the colonial time, present-day water users relying on Nduruma River must operate in a legal plural context that is made up of locally evolved water sharing practices and the water rights system crafted by the national government.

This chapter illustrates how new commercial estate owners (mostly international companies) in this sub-catchment must adapt to local concepts of legitimate water rights to survive. They must accept the fact that acquiring state water rights does not automatically translate into legitimacy at the local level. To understand the dynamic of the struggles for water rights in the sub-catchment we develop a conceptual framework based on Boelens echelons of water rights analysis (Zwarteveen et al. 2005, Boelens 2008) and the concept of legitimacy (Bodansky 1999, Bodansky 2007).

The chapter is organized as follows: Section 8.3 presents the conceptual framework used to explore water rights struggles in the Nduruma sub-catchment. Section 8.4 introduces the case study catchment. Section 8.58.4 presents the water governance context in the Nduruma sub-catchment. Section 8.5 further presents cases of rights struggles and negotiation processes, focusing on the role of large-scale irrigators in local water allocation systems. Section 8.6 is a discussion of the research findings in light of water rights theory and formalization. Finally, by way of conclusion, some lessons for the development of catchment water allocation systems are provided in section 8.7.

8.3   FRAMEWORK: WATER RIGHTS, STRUGGLES AND CONTROL

Property rights define an individual's rights, privileges and associated limitations of a specific resource use, and their allocation affects the efficiency of resource use (Schlager and Ostrom 1992). According to Bromley (1997), property provides some benefit streams, while a right to property offers security over that benefit stream. A property right defines a relationship between individuals (or groups) with respect to the use of a particular resource and the benefits this use generates. To have a property right is to have the capacity to require some authority system to defend your interest against the interest of others (Bromley 1997: 50). Because of its vital, rivalry and non-excludability characteristics, water is a unique resource, the management of which requires a suitable set of institutional arrangements. A water right is therefore often composed of set or "bundle" of graduated privileges that are assigned to different social entities (Schlager and Ostrom 1992, Shi 2006, Bruns 2007). It defines who is entitled to a certain amount of water, at a particular time and location, during scarcity.

The sources of water rights are multiple and dynamic. They often take many forms at many levels of water management. Water allocation is not necessarily a matter of formal licenses to abstract water or contractual commitments for water delivery, but also local understandings such as taking turn to use water, and when and where irrigation water may be used (Bruns 2007). According to Bruns (2007), real access to water depends on how water is allocated at multiple levels, among larger jurisdictions such as nations, states, provinces, and districts and among organizations and individuals extracting water from rivers and aquifers, as well as the crucial details of water distribution within irrigation systems. Water rights may be implicit in the design of structures, and asserted in decisions about guarding, maintaining, or modifying irrigation infrastructure (Bruns 2007, Lankford and Beale 2007). Even when water rights are formally stipulated, such entitlements must still be translated into seasonal and daily decisions about withdrawing water.

With increasing water scarcity, most governments find themselves walking a tight line. To create, maintain or restore order, governments assert full ownership of water, and in theory, also the sole authority to determine who is entitled to water at a particular point in time (Bruns and Meinzen-Dick 2001, Molle 2004). Water users (individual or groups) are granted 'official' licenses or permits to use a certain amount of water, at a particular location and duration. Domestic users may be granted access to water (mainly for drinking) without a permit. The duration of rights may be permanent, for a number of years, or made conditional upon productive use, but it does not usually take into account hydrological variability (Molle 2004). Common belief is that the state's prescriptive water right is legitimate and legal. However, alongside the statutory rights system are the local norms and customs that mediate day-to-day access to water.

By contrast, local water rights (based on customs and norms), are locally developed and adapted through step-by-step negotiation between the users - often building on pre-existing rights system. As described by Molle (2004), the process of negotiation occurs at several nested levels in a river basin:

1. At the river level, during dry seasons there may be insufficient flow to meet all demands and this gives way to negotiated rules for sharing between user groups (e.g., irrigation canals using one river), and they are constantly redefined.
2. Within an irrigation canal, users' participation in maintenance may be instrumental in the definition of allocation rules in case supply is unable to meet demand.

Locally negotiated water rights are often sanctioned by the authority vested in the decision-making body (e.g. River Committees) and by the social recognition of these structures. Also of importance is the fact that at the level of an irrigation canal, water rights are often tied to labor investment in the hydraulic property which enhances one's claim to water access (Coward 1986a).

Water rights claims are often contested. As depicted in Figure 8.1, the struggle includes competition over who gets access to water, infrastructure and material means (resources); contest over the formulation and contents of water rights and operational norms (rules); struggle over decision-making authority and the legitimacy of rights systems (regulatory control); and the diverging discourses that defend or challenge particular water policies, normative constructs and water hierarchies (regimes of representation) (Boelens 2008).

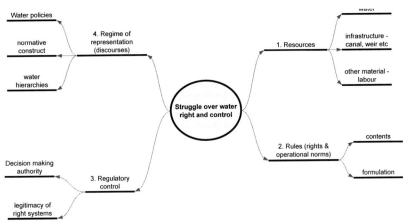

Figure 8.1: The proposed echelons of water rights analysis (Source: adapted from Boelens 2008).

The four components are involved simultaneously and chained together in particular ways, establishing how water is distributed, how humans and non-humans are ordered in socio-technical hierarchies, how this is legitimated by moral and symbolic order, etc. (Mehta 2007, Boelens 2008). Such alignments take place in ways that either strengthen or challenge the status quo. In this way, water rights struggles are at the centre of power relations. Power is used here to mean relational effects arising out of one's location advantage, access to other material resources and/ or psychological strength (Piccione and Razin 2009). Power relations generate key features of water rights content, distribution and legitimacy, and in turn, water rights in action reproduce or restructure power relations (Boelens 2008). For instance, acts of resistance against a dominant actor (e.g. attacks on estates intakes) may subvert power relations between the actors and this in turns affect water control.

In sum, water rights stand at the centre of struggles over legitimacy. Legitimacy describes the formal and informal ways in which processes, policies, structures and agents are validated and consequently empowered (Gearey and Jeffrey 2006). The challenge is in equating legitimacy to legality. Legitimacy has both a normative quality and a social dimension - it is not only a reason for action but also the justification for action (Gearey and Jeffrey 2006, Bodansky 2007). Bodansky (2007) argues that legitimacy is a much broader concept than legality in at least three ways: (1) legality is one possible justification of authority, but not the only criteria in assessing authority of institutions (e.g., a certain use of water may be illegal yet considered legitimate by some); (2) exercise of authority can exist outside a legal system and still raise issues of legitimacy (e.g. traditional leaders); and (3) legitimacy relates not simply to compliance, but to the justification of authority more generally. He argues further that an institution may be considered legitimate when users think that it serves their self-interest. We use the above conceptualization of water rights struggles to explore the dynamic of water access and distribution between large-scale irrigators and smallholder farmers in the Nduruma sub-catchment.

The complex nature of accessing water can be aptly illustrated by the case of Nduruma River. In this catchment small and large-scale irrigators mediate their differing perspectives because they share one and the same source of water.

8.4   RESEARCH METHODS AND CASE STUDY SUB-CATCHMENT

8.4.1   Research methods

The objectives of this research were to describe and analyze water-sharing agreements among and between various users of the Nduruma River. We were particularly interested in how each user's unique situation affected water use and cooperation with other users. Our consideration of situation included questions of: location (upstream or downstream); political influence; size of demand and/or production; type of intake and/or irrigation technology employed; and whether the user was foreign or local. To achieve these objectives, we first needed to identify all users, their location, and intake points. To identify the users and their networks, we followed the Nduruma River downstream, mapped the hydraulic infrastructures tapping the water source (predominantly irrigation canals referred to here as furrows), and followed them back to their owners.

Once each furrow was located and attributed to either a village or an estate, we conducted interviews with various stakeholders and irrigation committees. We started group interviews with village furrow committees: 7 members, including the village chairmen, each from Makasuro sub-village, Nduruma, Moivaro and Madawe villages; and 14 members from Manyire village. Discussions with these groups were conducted around a similar set of questions to determine current norms in use, how water is allocated to each furrow, and how this is transformed into water access for individual members of a furrow. We also investigated what the village leaders thought about how their water use influenced downstream users, and how their supply was affected by those upstream.

We also met with at least one representative from each of the estates located in the mid-zone of the Nduruma River: the owner of Old River Farm; the former owner of Gomba Estate; the farm manager of Dekker Bruins; the director and irrigation manager of Arusha Blooms; and the environmental and fertigation officers of Kiliflora. Questions asked to these individuals were similar to those asked of the villagers such as determining if their water supply met their demand and how they affected and/ or were affected by surrounding village users. In addition, representatives were asked to describe their relationship with surrounding villagers and their strategies to secure access to water.

Interviews were also conducted with officials from the local governments and basin authority to get a broader understanding of the issue: the Meru District Irrigation Officer; the Water Officer at the Pangani Basin Water Office; the Nduruma Ward Executive Officer; and the Sokon II Ward Office Chairman. We asked them to describe the current state of water sharing (or conflict) and how state-issued water rights were influencing the situation.

For those villages that had River Committees in place, we met and conducted interviews with the committees' head members as well. We held discussions with the chairman and secretary of the Nduruma and Manyire River committees. The committee members were asked to give their own perspective on the questions listed above. In addition, they were asked general questions about the committee itself: why it was formed and whose idea or initiative it was, how a furrow group can become a member of the River Committee, the specific role of this committee, especially with respect to water allocation, its spatial span of control, management structure and election of representatives and leaders, and its relationship with estates and estate managers.

8.4.2   Nduruma River

The Nduruma River, located in the upper parts of the Pangani River Basin, originates in a protected forest reserve near the summit of Mount Meru and is fed by small tributaries (Songota and Manyire being the main ones) and springs along its course (see Figure 5.4). Rainfall is bimodal with long rains (*masika*) from March to June and short rains (*vuli*) falling from November to January.

Eight administrative wards of the Arumeru district depend at least in part on the Nduruma River at some time in the year: Bangata, Nkoanrua, Sokon II, Mlangarini, Nduruma, Moshono, Kikwe, and Mbuguni. These wards can be divided into three groups based roughly on altitude and composition of the water users – highlands, midlands, and lowlands (Table 8.1).

The *highlands* begin below the forest reserve and end just above the Arusha-Moshi highway. Irrigation has been practiced for more than two hundred years in the highland zone. The villages in this zone are Midawe, Bangata, and Nkoanrua. The western bank of the river is mainly occupied by Arusha people while the eastern bank by the Meru people (Spear 1997). The main source of income for this area is the cultivation of maize and bananas, coffee, pyrethrum and round potatoes. The most upstream village of Midawe has several furrows that in fact draw from springs and the river Songota, a tributary of Nduruma. Off the Nduruma River itself, there are four main furrows. Due to the large population, land shortage is high in this zone, and this leads to degradation of the water sources and river banks, as people are forced to farm any available land.

Table 8.1: Agro-ecological zones of Nduruma sub-catchment.

Zone & wards	Elevation (AMSL*)	Rainfall (mm/yr)	Land use/dominant features
Forest reserve	Above 1800m	~ 1400	Forest reserve, wildlife, lumbering, mining. National park
Highland	1400– 1800m	~1000	Subsistence agriculture (major crops: coffee, bananas, sugarcane, yams, maize, vegetables, and sweet potatoes). Supplemental irrigation practiced but changes to full irrigation during dry seasons. Livestock are stalled – mainly dairy cattle. Water used for domestic, irrigation, livestock and Arusha municipal supply
Midland	1000 – 1400m	~500	Referred to as "Plastic Valley," the area is under intensive agriculture. Crops include coffee, banana, maize, beans, horticultural crops and export flowers. Originally white settlers' coffee estates but now changing into larger commercial flower growers (majority international investors). Competition over water is intense.
Lowland	Below 800m	~ 400	Subsistence agriculture and livestock. Receives low rainfall, highly affected by upstream water use. Major crops: tobacco, rice, beans, maize, and vegetables. Livestock are free-range, mainly own by pastoral Maasai.

The *midland zone* is roughly defined by the current Arusha-Moshi highway and the Old Arusha -Moshi dirt road. It overlaps with the area that locals call the "plastic-valley" because of the huge number of (plastic-roofed) greenhouses belonging to commercial farmers that dot the landscape. Formerly, the area was used for coffee production and formed what Spear (1997) called the "iron ring of land alienation" around the base of Mount Meru. The coffee estates around Mount Meru were not nationalized by the independent government of Tanzania (Spear 1997, Baffes 2003). These estates are foreign-owned and depend on Nduruma River for intensive irrigation to grow flowers and vegetable seeds needed for the international market. Smallholder farmers are the minority in the midlands, both in terms of their number and land holdings. Manyire tributary originates from this zone and is being used by the estates and surrounding villages.

The *lowland zone* of the Nduruma valley is semi-arid. Until recently, the area was used for commercial sisal production and livestock grazing. Currently, main economic activities include livestock keeping and subsistence agriculture. The majority of users are smallholder farmers who grow predominantly maize, beans, rice, tomatoes, vegetables and fruits. Some of the villages farthest downstream have been recently populated by people moving from different parts of the country. The most upstream portion of this zone also overlaps with the "plastic valley," and thus there is a fair number of large-scale irrigators here as well, clustered just below the Old Arusha-

Moshi road. The villages in this zone relying on Nduruma water are Mlangarini, Manyire, Mzimuni, Marurani, and Nduruma.

8.5   WATER GOVERNANCE IN NDURUMA

Irrigation along the Nduruma River dates back more than two hundred years (Spear 1997). But intensive water use only started to pick up pace during the colonial period, when white commercial farmers settled in the area, and began using the same water resources as the indigenous population. A map dating back to 1959 provides an illustration of the land use situation shortly before independence. This map was obtained from Pamoja archives; it is made by J.N.S. on 10-4-1959. Only the initials of the author are indicated. We think the map was made through a directive from the colonial administration (see Spear 1997). The majority of the furrows are in the highlands, and are labelled 'African.' Below them, in the midland there are fewer, but longer furrows, the majority of which are labelled 'European.' Below these (lowland), only a small number of short furrows can be seen, which are once again labelled 'African.' The situation is not much different today; the present geography is such that smallholder farmers are located upstream and downstream while the midstream zone is mainly occupied by large-scale irrigators growing flowers for European markets. This spatial geography shapes the nature of water governance in the catchment. This section briefly describes, first, the state-sanctioned water right governance system in Nduruma River, second, the local water governance arrangements, and third, the functioning of the locally developed Nduruma River Committee.

8.5.1   State-sanctioned water right governance in Nduruma

The present water rights system was initially introduced to curtail natives' water use and secure water for white commercial settlers who had interest in agricultural intensification in the highlands (see Chapter 2 section 2.5.3). In the Nduruma sub-catchment, local farmers were dispossessed of their lands and water resources (Spear 1997). Subsequent amendments have since been made to the Water Utilization (control and regulation) Act in 1948, 1959, 1974, 1981, 1996, 1997 and 2009. A volumetric water use fee was first introduced in the 1974 Water Utilization (control and regulation) Act. The Tanzanian government recently reformed its water sector: in 2002 a new National Water Policy was put in place (Tanzania 2002a). Under this policy, water belongs to the state, and all water users with an intention to abstract surface or underground water must acquire water rights from a designated basin water authority (basin water board/office). However, with the exception of volumetric water use fees, the current water rights system is still similar in many aspects to the colonial water law. A basin water board may grant or refuse water rights to any person or groups. If granted, the water right specifies the purpose, volumetric amount allowed, duration of the right, and the source (Tanzania 2009). A permit holder may, with the consent of the basin water board, temporarily lease his/her use right to anyone, and for any duration, provided that the duration of the permit is not exceeded. Nothing in any water use permit granted implies any guarantee that the quality and quantity of water referred to is, or shall be, available. Permits may be declared by the basin board as appurtenant to land described in the permit. However, state issued water rights seem to lead to over abstraction in many places. In the Rufiji basin, for example, issuing water rights led to over abstraction and increased competition as users argued that acquiring a state water right is synonymous to owning the water (see Van Koppen et al. 2007). The problem is that "water rights" was translated to Swahili as *haki za kumiliki maji* (water property rights) – which made users think that acquiring a permit conferred full ownership rights. The government tried to address this problem with the new Water Act (control and regulation 2009) by redefining the phrase "water use permit," using the Swahili *haki ya kutumia maji* (right to use water). In addition current water use permits are issued to users specifying a fixed flow rate, determined based on the assumption that an average supply exists (Lankford and Beale 2007, Van Koppen et al. 2007). However, in the Pangani and other basins in Tanzania rainfall and hence water supply is highly variable (Lankford and Beale 2007) which makes the state allocation inappropriate especially during low flows (dry seasons or during droughts).

The Pangani Basin Water Board (PBWO), through its executive basin office, is responsible for allocating all water rights in the basin in which Nduruma is a sub-catchment. PBWO maintains a database of all users in the basin and handles all new requests. Most of the large-scale irrigators along Nduruma acquired their water rights during the colonial time, but these were revised by PBWO in 2003 to accommodate Arusha City water demand. See Figure 5.5 for a sketch of water right status and measured dry season abstractions along Nduruma River.

Through their system of indirect rule, the British also created the conditions for a pluralist system of water governance in Tanzania (Spear 1997). They created Crown Land to be governed by statutory law and Native Reserves (land occupied by the Africans) to be governed by local law. Ever since the colonial time, customary rights in the Pangani River Basin have coevolved with statutory water rights.

8.5.2   Local water governance in Nduruma sub-catchment

The furrows in the *highlands* each have a committee, composed of a chairman, secretary, and members. For most of the furrow committees there is also a "council" of elders who act as advisors. The committees are mostly concerned with organizing the youth of the community into a maintenance schedule for the infrastructure. The furrow committees generally meet every three months and elections for membership are held every four years, which are held at the annual village committee meetings. In the dry season, when water is scarcer, there is a need to precisely allocate the water amongst the various farmers (via a rotational system). Most of the highland furrows drawing from the main stem of the Nduruma River have metal intake gates but furrows drawing from springs and Songota tributary have no lockable gates. However, all metal gates were found locked in the fully opened position, or intentionally destroyed in order to abstract more water, a sign that they are not used to regulate water allocation. All highland furrows abstract water simultaneously with no turn-taking enforced even during the dry season.

Highland furrow committees claimed that there is no conflict amongst the villages themselves. Although some downstream villages use these same furrows the highland furrow committees never organise meetings with them. Nkoanrua's furrow chairman explained that the reason why they have never met with downstream users such as Moivaro Village is because these villages tap springs which ensure domestic water supply. Because of this domestic supply, downstream villagers are believed to receive enough water to survive. The highlanders do not seem to bother whether or not the downstream villagers receive enough water to produce sufficient food and maintain their livelihoods.

In the *midland* area, only furrows used by smallholder farmers have committees with structures similar to those of the highlands. large-scale irrigators do not use committees to manage their furrows. During the rainy season, farmers in Moivaro for instance, claim that they don't irrigate so they only use intra-village allocation schedules in the dry season. In May and June, the furrow committee is in charge of repairing the furrows that were damaged, in preparation for the coming dry season. The furrows are typically silted up during the rains and the intakes get damaged. There is an annual fee of around 2,000 - 3,000 Tsh (1.34 - 2.00 USD) collected from each plot (*shamba*) in the village for completing the furrow repairs.

Lowland furrows have committees with the following composition: 3 water distributors, 3 advisors, 2 water guards, a chairman, treasurer, and secretary. These committees are responsible for water allocation to individual farmers, routine maintenance, and resolving resource conflict. The chairman, secretary, and advisors determine the day-to-day water allocation schedule. Individuals are assigned water for approximately two hours each. Despite this planning, villagers still steal water along the way. There is a 5,000 Tsh (3.34 USD) fine for water theft, which is handled within the village government system.

8.5.3   Local catchment wide governance structure: Nduruma River Committee

At the time of the research (2009) there was no overall water management institution for the Nduruma sub-catchment that was recognised by the the Pangani Basin Water Board[16]. This notwithstanding, the allocation of water between the midlands (mainly large-scale farmers) and lowlands was being managed by the Nduruma River Committee. The villages that actively participate in the River Committee are Mlangarini, Manyire, Mzimuni, Marurani, and Nduruma. According to the river committee chairman, in the past there was sufficient water but still during droughts

---

[16] A new river committee governing allocation between highlands and lowlands was created in July 2012 by the farmers.

or dry seasons, elders would meet and agree on allocation schedules. There was one elder whose role was to guard the river. He would follow the river upstream and negotiate with upstream farmers. But between 1962 and 1974, an extreme drought occurred and the idea of the committee emerged. Leaders of the individual furrows would meet after examining the levels of available water in the Nduruma River and then agree on allocation schedules. However, the discussion of allocation schedules in 'traditional' times never travelled beyond Ambureni/Moivaro Village to the highlands. The Nduruma River Committee in its current structure was initiated in 1999 by smallholder and large commercial farmers with support of the Arumeru District Commissioner. This is also the year when the first formal elections occurred for the chairman and secretary.

Currently, every furrow in the mid and lowlands is represented on the Nduruma River Committee, normally by the chairman and secretary of the furrow. Within each furrow there is an election every three years for these positions. If there is a problem, new members are selected to replace old ones. Among these representatives, a chairman and a secretary of the Board are elected. In addition, the security guards from each of the furrows attend the board meetings, but they do not vote. Representatives from each of the estates attend committee meetings as well – normally the estate farm managers or irrigation officers, who are always native Tanzanians. The current chairman of the River Committee is from Mlangarini Village and the secretary from Nduruma Village.

The Nduruma River Committee is responsible for setting the allocation schedules for each of the main villages as well as commercial estate furrows. If it is discovered that a furrow/estate is violating the agreement and abstracting outside of its allowed time, the board levies a fine on the responsible party. In the case of the estates, the fine is levied in the name of the estate representative to the board, usually the irrigation manager. Stephen Gregory from Tanzania Flowers explained that in his case the estate would cover the penalty. In the case of the village furrows, the fine is levied against the furrow chairman regardless of who made the irrigation offense.

Under local water-use bylaws, the punishment for stealing water was to supply a sheep or goat to be eaten by the clan. Since 1999 the fine has been levied in Tanzania shillings. The fine for estates was raised in January 2009 from 100,000 Tsh to 200,000 Tsh (67-134 USD). The fine for village farmers caught stealing is variable and usually a much smaller sum. This punishment is often relegated to the individual furrow committee to manage. In theory, if a villager is unable to pay his fine he is expected to forfeit a section of his land, but in practice this has never occurred.

There are no registration or membership fees for being a part of the River Committee. Most of the large estates make voluntary payments when asked. These cash allotments go toward small purchases such as refreshments for meetings. The River Committee has no bank account and thus no mechanism for storing large amounts of money. Nelson, from Dekker Bruin expressed his frustration concerning this matter. "Sometimes [the estates] are sent a request for funds, or someone asks for [help with] transport to the [meetings]. The board should really have an account.

There are silly problems like sometimes the chairman fails to phone because he has run out of credit." Because the requests for financial support are informal and sporadic, estates have no way of knowing if they are sharing the financial burden. It is apparent that Tanzania Flowers and Dekker Bruins have borne the brunt of the costs. The Nduruma River Committee has no headquarters or office and meetings are generally held in an open field on the property of Dekker Bruins.

Finally, Manyire villagers claim that they have a River Committee for Manyire river. The committee reportedly was elected in the year 2000 during a general assembly attended by the villages of Manyire, Maji moto, Karangai and Makasoru. Apparently Nambala and Kikwe village were not part of the meeting. Nambala village is in conflict with the other villages and does not recognise Manyire River Committee.

8.5.4   Legitimacy and struggles over water access and control

The situation in Nduruma is challenging for the large-scale irrigators that want to invest in an environment where the state's water law is deemed legal and legitimate at all levels. All large-scale irrigators in the sub-catchment have state-issued water rights (also referred to as 'official' water rights). Although some smallholder groups have also acquired state water rights on their irrigation canals, the allocation of water "on the ground" is being done according to local rules. This leads to struggles between the smallholder farmers, who appeal to customary principles and the large-scale irrigators, who want to adhere to the state's statutory water law. In addition to the struggle over water access and control, the large-scale irrigators also invest in water related infrastructure to secure access to water. The technological innovations include the use of high-tech drip irrigation system, rainwater harvesting from the greenhouse roofs, storage infrastructure and boreholes. We present in this section three cases of conflict and cooperation between the estates and smallholder farmers: first, Gomba Estate that claims that only the official legal right is legitimate; second, Enza Zaden that attempts to mediate conflict between Manyire users; and third, a group of estates that agrees on rotational allocation with smallholder farmers.

8.5.5   Contested official water law: case of Gomba estate

One notorious example of conflict between the Nduruma River Committee and a foreign-owned commercial farm is the case of Gomba Estate. In 1996, a Canadian investor took over a failed coffee estate in the midlands. The old coffee estate had two permanent water rights attached to the land: one water right is from Nduruma river (issued for Lambi 2 furrow) and another from nearby Manyire River (issued for Lambi 1 furrow). Lambi 2 was already being used by the village of Manyire. Gomba Estate embarked upon a large operation to grow a diversity of crops – mainly vegetables and fruit trees. The downstream villages of Manyire, Nduruma, and Mlangarini immediately noticed the decrease in dry-season water supply when the farm became operational.

On behalf of the downstream villagers, the Nduruma River Committee attempted to meet with the Canadian to negotiate a water-rationing schedule. According to the village leaders of Manyire and Nduruma, "He wouldn't attend any of the meetings to discuss water allocation. We tried to levy fines on him for taking water at the wrong time but he refused to pay and wouldn't let us in at the gate [of the estate]." The Canadian's reason for refusing to negotiate with the Nduruma Board was that he felt he had an "official" legal right to the amount he had been allocated by Pangani Basin Water Board/Office (PBWO) and for which he paid an annual fee. He also felt uncomfortable establishing agreements with an unofficial organization when it concerned the success of his estate. Explains the Canadian, "What you end up with are guards with machetes watching your water intake to make sure you don't open it too early. No one knows what is official or not. Inside the villages the villagers themselves steal the water – it is impossible to negotiate with that many people at once."

For support of his claim, the Canadian called the District Commissioner and cited his official water right. In his words, "All I would ask was, 'Please apply the law.'" Village leaders from Nduruma interpreted this behaviour as disrespectful and unaccommodating. The Canadian refused to solve anything without first calling the area Ward Councilor (*Diwani*). In the words of village chairman of Nduruma, "The Canadian only knows three people: the area member of parliament, the President, and the Minister of Investment. He was very rude; he wouldn't attend the River Committee meetings." In response to the Canadian's refusal to negotiate, many of the villagers responded with violence. Gomba's furrow intakes were vandalized and its irrigation workers were harassed. According to the Canadian, these were common occurrences during the dry season, "There were literally mobs of people with machetes at the intake from June to the end of February the following year." Gomba Estate stopped operation in 2007. The land of the estate now lies abandoned, but Manyire village regained control of Lambi 2 furrow.

8.5.6   Mediating local conflict: Enza Zaden's role in Manyire water conflict

Enza Zaden is a Dutch-owned vegetable seed breeding estate located along the Old Arusha-Moshi Road. In 2005, the company bought 45 acres of land that was once a coffee plantation known as "Sarkos's Farm". Although the estate has water rights to use both the Nduruma and Manyire rivers, it doesn't take any water from them because in the dry season there is not sufficient water. Instead, the main supply comes from a borehole near Lake Duluti, one kilometre upstream of the estate. This borehole has been registered with PBWO, for which the estate pays an annual fee. The estate has also drilled a second borehole for reasons of water security and claims to be in the process of registering it with PBWO. Enza employs 140 people on a permanent basis with an additional 40 temporarily hired to assist with the ongoing constructions on the farm. Almost all of these employees come from the downstream villages of Manyire and Nambala.

In early 2001 Nambala Village dug a new furrow from the Manyire River across the land now owned by Enza Zaden. At the time there was no agricultural activity taking place on this land. The furrow is called SANAKIMA, an acronym derived from the names of its users: Sarkos Farm, Nambala Village, Kikwe Village, and Maweni Village. Representatives from Nambala explained that this furrow was necessary because their main water source, the Ganana River, had dried up from overuse upstream. Nambala and Kikwe also claim to have been drawing water from Manyire River since 1978, at which time there were unofficial user agreements between the various village heads.

Downstream of Enza's estate and the SANAKIMA intake are the villages of Manyire, Karangai, and Maji Moto, with a total of sixteen furrows which use the Manyire River as their source. The seven furrows that support Manyire Village are: Majengo Juu, Majengo Kati, Mshikamano, King'ori, Kusini A, Kusini B, Levorosi, and Upendo. According to the chairman of the Manyire River Committee, representatives of the Board complained to PBWO about the construction of SANAKIMA furrow. He claims that PBWO, which does officially recognize the aforementioned seven furrows, ordered the closing of SANAKIMA. Meanwhile, Nambala's village committee claims that they are in the process of registering for an official water right with PBWO. A tenuous sharing agreement was agreed between the two parties concerning the times of opening and closing of the SANAKIMA furrow. However there continued to be many disagreements between the members of the villagers – each side of the argument sent a full-time watchman to guard the furrow intake point on Enza Zaden's property. On several occasions there were violent interactions involving machetes when one village accused another of either opening or closing the furrow at inappropriate times. The intake of SANAKIMA furrow was not fitted with cement lining or a control gate. This meant that water flow was controlled by the infilling of stones and soil excavated from the riverbank and from the nearby Enza Zaden farmland – a routine activity that contributed to the degradation of the source and the slow erosion of parts of Enza Zaden's land.

The managers of Enza Zaden became increasingly frustrated with the conflict situation. In addition to the harm being done to their land by trespassing villagers, the noise from the occasional brawl would wake up the manager and his family whose house is located near to the furrow intake. The manager of Enza Zaden met with representatives from each side and proposed that the intake point itself be moved farther downstream, away from the estate house. In addition, Enza Zaden proposed to fund the construction of an intake weir. A second meeting was held with water representatives from Manyire, Nambala, Kikwe, and Maweni in attendance. The estate's production manager facilitated the meeting. The previously agreed upon irrigation schedule was modified, put into a written contract, and signed on Enza Zaden official letterhead. While prior to the agreement Nambala, Kikwe, and Maweni received water from 2am to 3pm, the current (dry season) schedule has reduced their allocation to the period between 6am and 3pm. On the part of Enza Zaden, the company agreed to finance the construction of a permanent intake structure.

According to Enza Zaden's seed-cleaning manager, the villagers have so far stuck to the agreed furrow schedule; there have been no recent violent disputes over accusations of "intake gate" tampering. Throughout this process Enza Zaden made no attempt to contact any level of the local government or PBWO, though both Manyire and Nambala claimed to have requested intervention of the district commissioner without response. When asked what the purpose of signing the contract was, both parties replied similarly. The villagers think that it affords them some "evidence of agreement" against the other, though a local court is unlikely to consider this informal document binding. A member of the Manyire River Committee and resident of Manyire Village, claims that during a particularly heated encounter with Nambala furrow guards, he threatened to go to the police "with the signed document in hand." In response to this threat the men backed down. Villagers of Nambala, when asked for comment on the above anecdote explained that they understood that the document offered no real power of enforcement, "just the power of everyone signing." In return for its mediation role, Enza Zaden is treated like a relative by the villagers and many downstream farmers feel comfortable with the estate.

8.5.7   Negotiated allocation: Estates' agreeing with the local River Committee

During the rainy season (March - May and November - December) the villagers only use furrows for supplemental irrigation, but they engage in full-scale irrigation during the dry season. The estates are able to meet their full irrigation demands during rainy seasons and high flows. However, during the dry seasons, the Nduruma River Committee demands that all estates must reduce the time of their abstraction and the duration is negotiated every month. Most of the estates are active participants in the negotiation process and have representatives who attend every meeting. These estates include Arusha Blooms, Kiliflora, Tanzania Flowers, Dekker Bruins and Old River estates. The estates often contribute small amounts of financial support for drinks and transportation of members of the River Committees. Dekker Bruins particularly has invested in a strong relationship with the River Committee: all monthly meetings are held on the property of the estate and for each meeting Dekker provides refreshments for the participants. The estate's farm manager attributes this support to a necessity of cooperation. At the start of our research (January 2009), the arrangement stipulated that the estates and the village of Mlangarini abstracted water from 6am to 4pm. At 4pm their gates are closed and the water is allowed to flow downstream into the open furrows of Nduruma Village and the other users below the Maruroi furrow. This rotational agreement was revised in February 2009 and estates' schedule was from 5am to 1pm. To enforce this, the River Committee employs a water guard who patrols along the river, checking each and every intake of the large-scale irrigators. According to the director of Arusha Blooms, the agreement with the River Committee during the dry season means that they don't get enough water and the estates are forced to skip some irrigation schedules, which translate to an estimated 10% loss in production. She stated that over time, the problem is not water volume per se, but the lack of storage facility at the estate to maintain production during the dry season. The estates also negotiate between themselves. Arusha Blooms reported that they often send their representative (a Tanzanian national) to request additional time from the other estates and at times from the River Committee. The estate managers typically prefer to settle water concerns in this informal fashion without resorting to any outside authority (e.g., PBWO). Estates see their cooperative agreement with the downstream smallholder farmers as an act of good neighbourliness, although they also admit to trying to address complaints that could tarnish the companies' image internationally. The manager of Dekker Bruins stated explicitly during our interview, that to acquire a certification from the Netherlands they must be seen as working with the local communities. The manager of Tanzania Flowers stated that because one of its farms is located on the boundary between the large-scale irrigators and downstream smallholder farmers, they find themselves much more involved in the River Committee's activities, while the other estates further upstream interact very little with the board. The estates argue that their participation is an attempt to make downstream users understand that water use by the estate also generates benefits for the community downstream (e.g., employment opportunity, as well as schools and dispensaries constructed through their social responsibility projects).

However, the agreement is not without controversy. Some of the estates do not strictly abide by the River Committee's decision on water allocation, and Kiliflora is one such estate. Kiliflora uses an electric pump to convey water directly from the river to its reservoirs instead of employing the traditional gravity-facilitated furrow method. Unlike the other foreign-owned estates of similar size, Kiliflora irrigates using the relatively inefficient method of trough irrigation instead of drip. A trough is a U-shaped channel used to supply water and/or nutrient solutions to potted plants. According to the estate's fertigation officer, they run two pumps simultaneously, 6-8 hours a day seven days a week between 5am to 1pm. There are also two reservoirs located on the farms with storage capacities of 65,000 m^3 and 3,000 m^3. Also in this case the Nduruma River Committee determines the hours when Kiliflora is allowed to operate its pumps – during the dry season of January 2009 its allocation had just been reduced by an hour. Kiliflora's pumps are a matter of concern amongst the member of the River Committee and the downstream villagers whom they represent. In the fertigation officer's words, "We spent most of the last meeting negotiating over when we could operate our pumps. People from the village don't understand that we have a right to take this water." Representatives of the Board reported that Kiliflora often runs its pumps outside of its permitted time window and the irrigation managers of Arusha Blooms and Dekker Bruins corroborated this fact. However, the fertigation officer denies that Kiliflora has ever gone against the prescriptions of the Board. The fertigation officer however admitted that Kiliflora's withdraws more water than its PBWO-granted water right allows. "Sometimes we take more and sometimes we take less, in the dry season we take more than the water right grants." He justifies this trespass by explaining, "We do struggle with water…we struggle to maintain production during the dry season." The efficacy of the Nduruma River Committee was reportedly tested when the water guards found Kiliflora operating its water pump at unsanctioned hours. Kiliflora, like all large-scale irrigators in the area, is surrounded by a high fence and employs a security guard at its gates. When the

Board officials appeared at the estate to address this issue and collect the penalty, they weren't allowed in or granted an audience with the Kiliflora manager. Several letters explaining the infraction were ignored. The fertigation officer did not acknowledge that these events took place and insisted that his company follows the mandates of the Board.

It seems that, in general, the estates that participate in the Nduruma River Committee think that the organization is necessary for the region and are satisfied with its operation. The irrigation managers acknowledge that in the dry season water in the Nduruma River is insufficient to meet all demands. Dekker Bruins' representative explains that it is much better to have a venue for people to discuss their needs and complaints than to resort to violence:

> "There are more than 1,000 people downstream that need this same water. We cannot fight with them. In the past there was no discussion – machetes were always brought out whenever there were problems, but now the downstream villages are trying to organize."

A common complaint voiced by several of the large-scale irrigators is that there is a lack of continuous board activity throughout the whole year. They feel that the board is too concerned with matters of allocation and not enough concerned by longer term issues, such as source maintenance.

8.6 DISCUSSION

The Nduruma case illustrates the fact that sources of water rights are diverse, complex and often conflicting. It also highlights the issue that water rights do not arise solely from state laws, agencies and courts, but also from local institutions and views of other resource users. In Nduruma the question of legitimacy is at the centre of the water rights struggle. Water rights struggles in Nduruma conform to the four components of the water right analysis as proposed by Boelens (2008) but in a complex way. In this section we explore the dynamic of the Nduruma water right struggles. The order is for presentation only as the four components are involved simultaneously.

First, the competition over possession and use of resources in the sub-catchment dates back to colonial time. As described by Spear (1997), land ownership was a heavily contested matter by local Meru communities, colonial administration and European settlers throughout colonial time. The German and British colonialists alienated land around the base of Mount Meru. The land and labour control struggle was only brought to a close when the Meru people protested and appealed to the United Nations against British seizure of more land (see Spear 1997). The induced land inequities however still shape present days struggles over who gets access to water; the competition is at its most intense between the large-scale irrigators(midland) and the downstream smallholder farmers. In their midland position, the estates are sandwiched between smallholders farmers who are located both upstream and downstream of the sub-catchment. Hence, the estates are both advantaged and disadvantaged in term of hydraulic location.

At the second level is the contest over the content of water rights and its enforcement. Large-scale irrigators have location advantage over the downstream farmers, and from their international origin, they also have better access to other resources (e.g., knowledge, better irrigation technology and financial means). Estates also benefit from a close connection with a national government which is interested in encouraging foreign investment. The estates also claim water access based on state-issued water rights, which are labelled "official" and therefore legitimate. Hence in their midstream location, the estates may be considered more powerful. The estates water allocation, especially during dry seasons, does not go unchallenged by the downstream smallholder farmers. These farmers demanded that allocation should be rotational and take into account supply variability and not absolute values specified in the government water right. Most estates in the midstream engage in negotiations with the smallholder farmers and tend to agree on a time-based allocation during scarcity even when this implies loss of production.

This research finds that the most successful large-scale irrigators are those able to engage with local systems of negotiation and rotational water allocations. By adopting this strategy, the estates do not only avoid conflict with the local farmers, but also gain social reputation in the area, increasing the chance of cooperation from the farmers towards their hydraulic infrastructure investments. The main reasons for negotiation, according to their representatives, are that it is: (1) much better to have a venue for people to discuss and complain than to resort to violence; (2) a matter of water security; and (3) a organized front that can forcefully represent their interest at larger scale (catchment wide or to the Pangani Basin Water Office). Their cooperation seems to fit the argument that the larger one's stake, the larger one's interest in the common good and thus the more responsibly one may act (Van der Zaag 2007). Violent response by the smallholder farmers on estates' water infrastructure is sufficient incentive for the estates to cooperate and broker cooperative and equitable deals. This is illustrated by the case of Enza Zaden Estate stepping in to mediate water conflict between two villages. In return for its help, it gains respect of the villages. Gomba Estate, by contrast, provides a situation where non-cooperative behaviour led to self-destruction in the long run, as the estate was forced out of business with lack of water being a major factor.

At the third level, regulatory control, the chapter has shown that this power is largely exercised by the Nduruma River Committee, which claims to have legitimate decision-making authority and takes responsibility for allocating water between midland and lowland farmers. This cooperative arrangement seems to only work well between close neighbours; at larger spatial distance it is less effective. The major weakness of the Nduruma River Committee is that its membership only encompasses large-scale irrigators and the downstream users. The board is simply managing water the upstream villages were unable to use. Since highland farmers do not have River Committees, it is difficult for the Nduruma River Committee to engage with them.

The fourth level dealing with regime of representation is best described by the water management perspectives of the different levels of government. The national government is interested in enforcing statutory water policies, laws and hierarchies, as well as promoting foreign commercial investment. However, the local district administrations are more likely to spend their limited resources "keeping the peace," rather than enforcing the letter of the water law. Estates that refuse to cooperate with smallholder farmers are often told by district administrators to go back and negotiate with their neighbours.

8.7 Conclusions

This chapter has described how water access is negotiated between smallholder irrigators and large-scale irrigators sharing the water of Nduruma River, Upper Pangani River Basin, Tanzania. The spatial geography of Nduruma is such that smallholder farmers are located upstream and downstream, while large-scale irrigators are in the midstream part of the sub-catchment. There is not enough water in the river to satisfy all demands. The majority of the smallholder farmers currently access water under local arrangements, while large-scale irrigators have obtained state-issued water use permits.

Although in such a context one would expect the weaker downstream farmers to lose out, instead cooperation prevails. Smallholder farmers in the sub-catchment counter inequities in land and water distribution by enforcing suitable allocation proxies (proportional division, time-based turns) which make water rights more meaningful. Powerful estates that do not agree to the terms of the local agreements find it difficult to keep on operating, as their water infrastructure may be vandalised by smallholder irrigators. In addition, local government officials pay little attention to the pleas by estates, thereby leaving the smallholder farmers with sufficient autonomy (see also: Schlager and Ostrom 1992).

The case study shows that a River Committee, an institution that was locally established and engineered and that is not formally recognised in statute law in Tanzania, has been conducive in structuring water allocation in a manner that has been effective and that has forestalled major conflicts. The institutional form of a River Committee has thus bridged local rules and statute law with respect to water.

We find the most successful estates are those able to cooperate with the smallholder farmers. The risk of the estates to lose a lot is here sufficient incentive for them to cooperate and broker cooperative and equitable deals with their less powerful counterparts (cf. Baland and Platteau 1999). In such a situation it becomes less easy for the many small water users to defect. But cooperative behaviour by the estates may also be due to other interdependencies between them and the smallholder farmers. Large-scale irrigators have to engage with their downstream villages because of their dependence on local labour. In addition, cooperation with smallholder farmers helps reinforce the water claim of the estates at larger spatial scales. However, this case study shows that at the larger catchment scale it has so far been impossible to institute and maintain effective cooperative arrangements, despite the formally established structures.

PART 4: EVOLVING WATER INSTITUTIONS:

DISCUSSION AND CONCLUSIONS

"If you didn't grow it, you didn't explain it" - Joshua Epstein in Generative social science

This section attempts to address the third sub-objective: develop a game theoretic model for considering alternative scenarios for collective catchment management (Chapter 9). It will also address the overarching research objective of studying conditions for reconciling state-led institutional arrangements and local water management practices (Chapter 10).

It has been demonstrated that a certain level of inequality interact with water asymmetry and heterogeneity to sustain collective action (Chapter 7). Can we reproduce this kind of solidarity-based water sharing arrangements? Epstein (2006) argues that to be able to explain such phenomena you have to be able to reproduce it - meaning generate the observed phenomena in a computer simulation. According to Epstein (2006), if the distributed interactions of heterogeneous agents can't reproduce it, then we haven't explained its emergence (Epstein 2006). Initially, the objective of this research was to complement empirical findings with agent based modelling (ABM). However this was not possible within the timeframe of the PhD research. Alternatively, an attempt has been made using a simple game theoretic model of irrigation canal maintenance to understand the emergence and functioning of these various sharing arrangements and alternative scenarios for collective catchment management (Chapter 9). Chapter 9 thus provides an initial attempt to model the emergence of self-governing systems from the empirical material collected.

Hingilili farmers playing the River Basin Game

Chapter 9

A GAME THEORETIC ANALYSIS OF EVOLUTION OF

COOPERATION IN SMALL-SCALE IRRIGATION CANAL

SYSTEM[17]

9.1 ABSTRACT

Local self-governance arrangements can often solve water allocation challenges. It is increasingly argued that innovative water governance arrangements should be those that build on the success of those long-enduring water-sharing arrangements that locally evolved. However, before practices can be scaled up, it is important to understand why they emerged, how they function, and can be sustained. Based on in-depth field research in the Pangani river basin Tanzania, we described several local water allocation arrangements that have evolved over the past 50 to 200 years (Chapter 6-8). In many of these irrigation systems, there is significant inequality of endowment (e.g. access to land). Despite this relatively high disparity between the actors, the irrigation infrastructures are being sustained. At the level of a river, we find a diversity of water-allocation arrangements that have arisen in the same river basin. In this chapter, we use a simple game theoretic model of irrigation canal

---

[17] Based on Komakech et al. 2012. Paper prepared for *the International Conference on Fresh Water Governance for Sustainable Development*, Drakensburg Sports Resort, South Africa 4-7 November 2012.

maintenance as a first attempt to begin to formalize the emergence and functioning of these various water sharing arrangements. Using the game we attempt to improve our understanding of how these arrangements arose and could endure, given the inequality in landholding, location asymmetry, and differences in the costs and benefits of collective action.

9.2 INTRODUCTION

Can we explain why some self-organised water institutions endure over time while others do not? As a common pool resource (CPR), the effective allocation of water requires deliberate coordination. Specifically, it is difficult—though not necessarily impossible—to exclude members from using the resource. In addition, water use by one appropriator reduces the amount available to the next potential user. How do effective allocation schemes arise in such a world?

In a river (or irrigation canal) where water flow is unidirectional, it is difficult to promote collective action. This is because the fixed location of users along the river imposes asymmetries, since upstream water use impacts downstream interests but not vice versa. This asymmetry may impact the willingness of participants to participate in collective activities such as providing labour for maintenance. However, interdependence[18] arising from differences among the appropriators has been identified as a factor that may counterbalance the effect of location asymmetry (Komakech et al. 2012d).

This may include differences in soil types, micro climate, crops, and social-cultural and kinship relationships that exist among irrigators in different locations (Komakech et al. 2012d). Lansing and Miller (2005) present a form of ecological interdependence where the choices of upstream farmers are linked to those of the downstream ones via the spread of pests and diseases affecting the entire ecosystem. Interdependence may also arise when labour requirements for canal repairs is high, e.g. rebuilding river diversion or clearing first stretches of canal that may cut through rough terrain before reaching the command area. Scholars suggest that under asymmetric water access, tail-enders often reduce their investment in infrastructure maintenance if they don't receive a fair share of the water produced (Janssen et al. 2011a). The fact that downstream may reduce their labour contribution if they do not get a fair share of the water could be sufficient incentives for the upstream farmers to cooperate and allow downstream farmers access to water. Janssen et al. (2011a) suggest that since the tail-enders know beforehand that they are disadvantaged they expect water scarcity and plan for it. They argue that as long as the system is perceived as fair by downstreamers the irrigation canal is likely to be maintained and will survive over time (Janssen et al. 2011a). In such cases the upstream irrigators need the cooperation of their downstream counterparts. In addition it is reported that costly

---

[18] Interdependence is defined as the mutual dependence of two or more actors on one another - it is the biophysical, social, and economic feedback effects of an actor's inactions or actions.

punishment (Boyd et al. 2010, Janssen et al. 2010, Sääksvuori et al. 2011) and/or communication (Hackett et al. 1994) increases the level of cooperation among the resource appropriators.

Although the above may be true, it omits the essential role of inequality and other differences between the actors that could be sustaining collective action. First, there is inequality in access to irrigation land among the irrigators. In one irrigation canal in Makanya catchment, Tanzania, Komakech et al. (2012c) found that the largest 20% of the farmers have access to 50% of the irrigation area and that the smallest 50% of the farmers control about 20% of the land area, yielding a Gini coefficient of about 0.58. Although there is relatively high inequality with respect to access to land and thus water, the Makanya irrigation canals are being maintained and have been in operation for more than 50 years. Second, farmers do have access to multiple lands in different locations of the irrigation canal and this may impact on their participation in irrigation maintenance and the way water is shared in the system. These differences may interact with other factors such as kinship to counterbalance the negative impact of asymmetry arising out of the unidirectional flow of water. It is important to investigate how these factors (e.g. inequality of land endowment, multiple land ownership, and other differences) contribute to collective action in an irrigation canal. This is the basis of the simple canal cleaning game we describe in this chapter.

Significant experiments have been conducted on CPR dilemmas where the resource users occupy symmetric positions with respect to the resource (see Baland and Platteau 1999). Although these studies fit well with CPR such as groundwater, fisheries, and forests, it does not cover the asymmetric access situation found in rivers and irrigation canals. We attempt to address this using a simple canal cleaning game which incorporates the situation of water asymmetry found in irrigation canals and rivers (see also Ostrom and Gardner 1993, Budescu and Au 2002, Lankford et al. 2004, Lansing and Miller 2005, Cardenas et al. 2011, Janssen et al. 2011a).

We incorporate land access inequality in the simple canal cleaning game by allowing some farmers to own more profitable land and/ or have multiple lands in different location. In this game, the irrigation canal is divided into sections. To grow a crop farmers must clear the canal of debris and rebuild its river diversion. Each period, farmers independently decide on the canal sections they will contribute labour for clearing. Their decision determines the amount of payoff each farmer receives.

The canal cleaning game is based on field observations of long enduring irrigation systems in the Pangani river basin, Tanzania. We observed that during irrigation canal repair, the work is shared evenly among those who are present at the site. Using game theoretical analysis we explore the impact of having costly canal sections, differential benefits from crop and multiple land ownership.

The chapter is organized as follows: Section 9.3 describes the setup of the canal cleaning game. Section 9.4 presents the analysis of the different configuration of the game. Section 9.5 gives a discussion of the results in light of empirical observation.

9.3   THE CANAL CLEANING GAME SET UP

The canal cleaning game consists of N-players (farmers) who own land along an irrigation canal. The locations of the farmers are fixed along the canal divided into n-canal sections (n>1 and N<=n). We first describe here a 2-farmers system (Figure 9.1) and then explore in detail a 3-farmers system in the subsequent section.

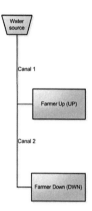

Figure 9.1: Schematic representation of the 2- farmers' canal cleaning game.

To grow crops the upstream farmer requires the first canal to be cleaned, while the downstream farmer needs both canals cleaned. If a clear canal is not present, then the farmer will receive a payoff of 0 from his/her crop (i.e. crops are totally dependent on irrigation). Crop yield is normalised to 1 unit if a farmer gets water, zero otherwise (the payoff reflects the value of the crop minus any agronomic costs incurred during the crop growing period). Each cropping season, farmers independently decide on which section of the irrigation canal they will clean (e.g. none, the first canal, second canal, or both canals). Assuming the cost of cleaning the first canal is " φ " and the cost of cleaning the second canal is "β" and that the cost is shared evenly between whoever participates (e.g. a farmer cleaning the second canal section alone contributes β, or if they work together, they both contribute β /2). We assume that 0 < φ <1 , 0< β < 1, because if φ >1 and β >1 then it is never worth it for a farmer to grow crop, as the cleaning cost exceeds the maximum value a farmer could derive from his or her crop. In this game a farmer's interest is to maximise his or her benefits from the crop. For notation we use a pair of bits [0, 1], where the first bit is equal to "1" if the first canal is cleaned and "0" otherwise. The second bit is similarly notated for the second canal (e.g. a farmer cleaning the first canal only is notated as [1, 0]). We notate farmer choices as [00; 00], the first pair of bits in the square bracket correspond to the downstream farmer and second pair correspond to choices of the upstream farmer. This notation of choices follows game theory convention, whereby the payoff (in the case above indicate decisions) by the row player (here, downstream farmer) is given first followed by the payoff of the column player (here, upstream farmer). The payoff bi-matrix for a normal form game is constructed as shown in Table 9.1. The payoff of the downstream farmer is the first value in each of the cells.

Table 9.1: Payoff bi-matrix for a 2-farmer canal cleaning game.

		Upstream farmer			
		None [0,0]	First [1,0]	Second [0,1]	Both [1,1]
Downstream farmer	None [0,0]	0 , 0	0 , 1- φ	0 , - β	1 , 1- φ -β
	First [1,0]	- φ , 1	-0.5 φ , 1-0.5 φ	1- φ , 1- β	1-0.5 φ , 1- 0.5φ -β
	Second [0,1]	- β , 0	1- β , 1- φ	-0.5β , -0.5β	1-0.5β , 1- φ -0.5β
	Both [1,1]	1- φ -β , 1	1- 0.5 φ -β , 1-0.5 φ	1- φ -0.5β , 1-0.5β	1- 0.5 φ -0.5β, 1- 0.5 φ - 0.5β

From the basic model we can derive four additional configurations by: 1) introducing relatively costly canal segments and head-works; 2) having different values for the upstream and downstream crops to highlight potential differences in soil and/or land sizes; 3) adding more farmers and land into the setup; and 4) allowing for multiple land ownership, that is farmers may own two or more irrigated plots in different parts of the canal. In the current game farmers do not take turns extracting water. It is assumed that once the canal is cleared there is sufficient water for all farmers.

9.4   MODEL RESULTS AND ANALYSIS

Analysis of the basic setup (2-farmers system) yields interesting insights into the system when farmers strive to maximise self payoffs or social benefits. For $\varphi = \beta$, we arrive at the following equilibria. If farmers were to maximise their payoffs we find the following Nash equilibrium for the 2-farmer set up (i.e. equal canal cleaning cost "β" and same crop value from the land). For $0.5 <= \beta < 1$, a unique Nash equilibrium exists where both canals are cleaned, with the upstream farmer cleaning the first canal and the downstream farmer cleaning the second canal [10; 01]. The intuition is that if it costs less than the value of the crop to clean one canal section but more than the value of the crop to clean all the sections alone it is worth it for the farmers to clean and grow crop. In this case, the downstream farmer can not clean both canals alone as the total cost will exceed his/her benefit from the crop. There are however multiple equilibria when $0 < \beta < 0.5$. In addition to the equilibrium state where each farmer cleans their canal section, another Nash Equilibrium exists in which the upstream farmer free rides on the downstream farmer [11; 00]. This is possible because the cost of cleaning one canal is less than half the value of the crop, so it is still worth for the downstream farmer to clean both canals and still make some profit even if he does all the work alone.

In situations where farmers would coordinate their actions the 2-farmer game set up yields many social equilibria for canal cleaning cost in the range $0 < \beta < 1$ (i.e. all the situations where both canals are cleared regardless of who clears them producing a surplus of 2-2β). That is if β <1, it is always better to farm than not, as long as you only have to clear one canal per plot. How these canals get cleared (whether it is by one or two farmers) does not matter as the costs just get shared among the workers (which does determine the share of the 2-2β surplus that goes to each farmer, but it does not impact the total cost of clearing).

In Figure 9.2, we first explored the parameter space for the setting where the upstream canal is more costly to clean but the benefit from crops is the same for both farmers. Figure 2 shows the parameter space for differential cleaning costs φ and β for the first and second canal respectively.

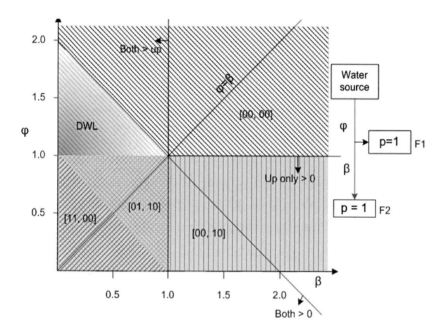

Figure 9.2: Parameter space for canals with differential cleaning cost. Upstream section cost "φ" and downstream section cost "β" to clean respectively. Benefit from crop "p" is the same for all land and equals 1. (DWL= Deadweight loss).

The following equilibria can be derived for different values of "φ" and "β" when the farmers maximise self-benefits (also indicated in Figure 9.2):

- [00; 00] No farmer cleans if φ >=1

- [11; 00] Downstream farmer will clean both canals while upstream farmer free rides if the total cost of cleaning both canal sections is less than the value of crop in the downstream plot, i.e. φ + β<=1

- [00; 10] Only the upstream farmer cleans his/her section if φ <=1 and β>=1

- [10; 01] Both farmers clean their section if φ <=1 and β<=1

The global solution is: cleaning only the upstream is beneficial if φ <1; cleaning both sections is beneficial if φ + β<2; and cleaning both canal sections is better than cleaning only the upstream canal section if β<1. When the cleaning costs are in the range 1 < φ <2 and 0 < β <1 (e.g. having costly intake), it is still possible for the farmers to make some profit if the farmers coordinate their actions. Without coordination however, no canal gets cleaned. We indicate this situation as deadweight loss (DWL) in Figure 9.2. The upstream farmer can not clean the first canal alone, it is too expensive, and it is also too expensive for the downstream to clean the entire canal system alone. If the farmers can negotiate a binding contract, e.g. agree to coordinate their effort, surplus would still be produced which can be shared among the farmers.

Next we explored the parameters space for the set up in which the benefits from the crops differ between upstream and downstream but the cleaning costs are the same for both canals; i.e. φ = β (Table 9.2 present the payoffs bi-matrix); assuming the crop yield in the upstream plot is "x" and in the downstream plot is held constant as the unit value of 1 as indicated in Figure 9.3.

Table 9.2: Payoff bi-matrix for a 2-farmer canal cleaning game with differential benefits.

		Upstream farmer			
		None [0,0]	First [1,0]	Second [0,1]	Both [1,1]
Downstream farmer	None [0,0]	0 , 0	0 , x - β	0 , - β	1 , x - 2β
	First [1,0]	- β , x	-0.5 β , γ -0.5β	1- β , x - β	1-0.5β , x - 1.5β
	Second [0,1]	- β , 0	1- β , x - β	-0.5β , -0.5β	1-0.5β , x - 1.5β
	Both [1,1]	1- 2β , x	1- 1.5β , x -0.5 β	1- 1.5β , x -0.5β	1- 0.5β , x -0.5β

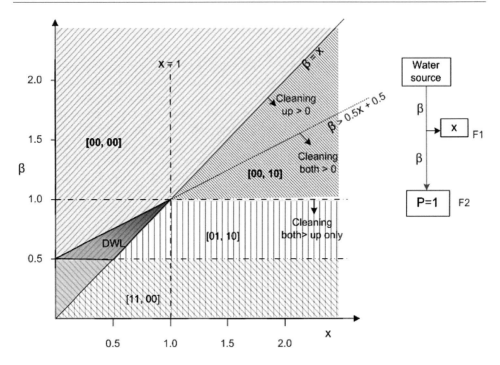

Figure 9.3: Canal setup with differential benefit from crop but equal canal cleaning costs.

We can identify the following equilibrium states for different values of crop yield "x" in the first plot and the equal cleaning cost per canal section "β":

- [00; 00] No farmer cleans if β>=0.5 and x <=β
- [11; 00] the downstream farmer will clean both canals while the upstream farmer free rides if β <= 0.5 and x<=β
- [00; 10] the upstream clean his/her part while downstream farmer does not clean if β>=1 and x>=β
- [10; 01] Both farmers clean their section if β<=1 and x>=β

A global solution is achieved in which cleaning the upstream canal is only profitable when x > β and cleaning both canals is only profitable if x +1 > 2β. Cleaning both canals is better than cleaning only the upstream canal if x > β and β<1. Through coordination, it is possible for the farmers to produce a surplus even when 0.5< β <1 and 0.5< x <1.

In a 3-farmers setup (see Figure 9.4, farmers F1, F2 and F3, with farmer F1 upstream, F2 midstream and F3 located further downstream) we explore three possible configurations for equal crop values (a = b = c=1): 1) where φ = β =γ; 2) where φ = 2β =2γ; 3) where F1 also owns the third plot (F3 = F1); and here (F2 also owns the third plot (F3 = F2).

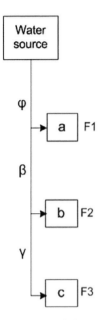

Figure 9.4: Schematic representation of the 3- farmers' canal cleaning game.

In Table 9.3 the first group of bits are choices of F1, the second group belongs to F2 and the third group belong to F3. We analyse three different cases as indicated in Table 9.3. Case 1 is a setup where there are three farmers (F1, F2 and F3) each having one plot, in the second case (Case 2) there are two farmers but one owns two plots (indicated as F1, F2, and (F3=F1)), while the third case (Case 3) is a setup similar to the second case but here it is F2 that owns two plot (indicated as F1, F2, and (F3=F2)).

Case 1: When φ =β =γ and φ <= 0.5, the canal always is maintained although some of the farmers will free-ride (e.g. F1 free-ride while F2 and F3 cleans with one of the farmers cleaning at least two plots). There are five possible equilibria when φ <0.5. The first three is where F1 free rides on the others, while F2 and F3 clean all canals (Table 9.3). The fourth equilibrium is where F2 free rides while F1 cleans the first canal and F3 cleans the second and third canal. The fifth equilibrium is when each farmer cleans his or her section of the canal. When the cleaning cost is in the range 0.5< φ <1, an equilibrium exists such that farmers clean their part of the canal only. The 3-farmers' setup is in many ways similar to the 2-farmer system. When φ =2β =2γ as long as φ <1.0 the canal is still maintained even if one farmer free rides on the others. However, the potential for free riding increases with an increase in the number of farmers.

Table 9.3: Possible equilibrium states for different canal cleaning costs and setup for the three plot system.

Canal cost	Range	Case 1 [F1 F2 F3]	Case 2 [F1 F2 (F3 = F1)]	Case 3 [F1 F2 (F3=F2)]
$\varphi = \beta = \gamma$	$\varphi <= 0.5$	[000, 010, 101]; [000, 110, 001]; [000, 100, 011]; [100, 000, 011]; and [100, 010, 001]	[111, 000]; [001, 110] and [101,010]	[000,111]; and [100, 011]
	$0.5 < \varphi < 1.0$	[100, 010, 001]	[101,010]	[000,111]; and [100, 011]
	$\varphi > 1$	[000, 000, 000]	[000, 000]	[000, 000]
$\varphi = 2\beta = 2\gamma$	$\varphi < 0.5$	[000,100, 0011]; [100, 010, 001]; and [100, 000, 011]	[011, 100]; [001, 110]; [111, 000]; and [101, 010]	[000,111]; [100, 011]
	$0.5 < = \varphi <= 1.0$	[000,100, 0011]; [100, 010, 001]; and [100, 000, 011]	[011, 100]; [001, 110]; [111, 000]; and [101, 010]	[000,111]; [100, 011]
	$1 < \varphi < 1.5$	[000, 000, 000]	[000, 000, 000]	[000, 000, 000]

Case 2: When $\varphi = \beta = \gamma$ there are multiple equilibria for $0 < \varphi < 1.0$. If $\varphi < 0.5$ there are three possible equilibria: 1) F2 free rides while F1 (the farmer with more land) does all the work; 2) F1 farmer cleans the first canal, F2 cleans all two; and 3) F1 cleans the first and third canal, F2 cleans only the second canal. We get only one equilibrium when $0.5 <= \beta < 1$ where each farmer cleans the section leading to their land. When $\varphi = 2\beta = 2\gamma$ as long as $\varphi < 1.0$ the canal is still maintained even if one farmer free rides on the others.

The third case (*Case 3*) is mirror of the second case. It does not provide new insights only that it is now the downstream farmer doing most of the work, while the upstream farmer can choose to free-ride.

Overall from this setup we can conclude that the irrigation canal will always get maintained even if some farmers put more efforts than others. More importantly it is the farmers with more plots that are likely to put in more effort. This needs to be verified with real farmers playing the canal cleaning game. The simple game however shows that this is likely to fail when it is too costly but we observe in real field settings the canals are still being maintained. This is because there is no unidirectional causal relationship in the simple canal game. The canal game assumes that once the canals are maintained water is available to all the farmers this is a simplification of the irrigation system.

9.5 DISCUSSION AND CONCLUSION

Common pool resources in which the appropriators are located in series and have to sequentially access the resources are likely to lead to inequity in distribution of the resource units - upstream users in many instances exploit the water asymmetry and appropriate larger share of the water produced (Ostrom and Gardner 1993, Janssen et al. 2011a). Based on field and lab experiments, scholars have concluded that in situations where upstream users appropriate a very large share of the resource relative to downstream shares the common resource will be underprovided (Janssen et al. 2010, Cardenas et al. 2011). If the tail-enders do not get a fair share of the resource from upstream, they may reduce their contributions to maintenance lowering the efficiency of the infrastructure in the long run (Janssen et al. 2011b). Hence in irrigation systems where upstream farmers are dependent on the labour of other members, upstream users balance their water use with downstream perceptions of fairness and equity with respect to access to the resource. This may put an upper boundary to the inequality as for example was observed in Makanya catchment (Chapter 6, Figure 6.3).

The analysis of the simple canal game provides some insights for the situation where all irrigators were equal in terms of access to land area and crop productivity: a) multiple equilibria exist in the system if it costs less than half of the expected yield from crop to clean one section of an irrigation canal. In this case it is sometimes worth it for downstream farmers to put in more effort in case the upstream farmer free rides; b) the canals will be maintained as long as the cost per canal section is less than the value of the crop to a farmer. However, for costly segments (e.g. high labour requirement for building river diversions) the irrigation infrastructure can only be maintained when farmers coordinate their efforts or strive to maximise the group benefits. Although surplus can still be produced, rational self-interested farmers in this game will not contribute labour when they are required to invest more than the value of crop to clean a canal. The need for coordination is even higher with an increase in the number of farmers, if benefits are differentially distributed among the irrigators, having more costly canal sections and an increase of the number of plots.

The simple game theoretic model shows that collective action would not emerge when actors maximise their self-benefit. Our model does not include situations where actors can come together to make binding rules. In other words, the model does not include the effect of power relations, kinship and other networks on the dynamic of water allocation between the irrigators. In field settings, farmers do share the burden of clearing canals more or less equally but actual water allocation is often contested with powerful farmers often getting more water than others (Komakech et al. 2012d).

In the three-farmer setup it is the farmer owning two plots who cleans more than one canal. This highlights the role of inequality in promoting collective action. The game predicts that for farmers maximising self-benefits the canal would not be maintained if the head section costs more than the crop to a farmer and there is no coordination. However, to be able to simulate the effect of unidirectional flow of water, we will have to introduce river discharge in the game, allow farmers to take turns extracting water after canals have been cleaned, and put a number of irrigation canals in series.

The simple canal cleaning game is not able to explain why some self-organised water institutions endure over time while others do not. However, it provides some insight into how small-scale irrigation systems may function generally. As farmers recognise that they are dependent on the cooperation of others for their survival they will more likely agree to contribute to collective action. The simple canal game predicts that no rational farmer would be willing to provide more labour for maintenance if the cleaning cost of some section exceeds the potential benefits. This simple game theoretic analysis is presented here as a proof of concept, further research is needed to simulate situations of water asymmetry, inequality and heterogeneity. The concept will be tested if played with real farmers.

Chapter 10

DISCUSSION AND CONCLUSIONS: THE EMERGENCE

AND EVOLUTION OF WATER INSTITUTIONS

Increasing water demands and other intervening phenomena such as increased climate variability in most catchments dictate the need for institutional arrangements that can improve equitable and sustainable use of the limited water resources. This thesis explored and analysed water governance processes as undertaken by the government and resource users at the local level and catchment scale. The overarching research objective was to study conditions for reconciling state-led institutional arrangements and local water management practices. The research used several concepts and theories to achieve the above objective. Particularly, the thesis demonstrated the importance of understanding how local level institutional arrangements emerged, how they evolved over time and how they interact with national policies and basin-wide institutions that have been established more recently.

In section 10.1 I will address each of the three research sub-objectives. In section 10.2 I synthesise the contribution this thesis has made to theories, concepts and methodology. In the final section (10.3) I present a critical reflection on the research.

10.1 UNDERSTANDING THE DYNAMICS OF WATER INSTITUTIONS IN THE PANGANI

The first sub-objective: attempt to understand the impacts of state intervention in catchment water management and its interaction with local water management norms and practices.

The state-led formalisation of water allocation and management in the Pangani basin, Tanzania, has so far had little of the intended impact on actual day-to-day water allocation practices. Although it is widely considered that allocating water rights or use permits would in water stressed catchments improve equity and reduce conflict, the findings in this thesis indicate that the 'paper' based water rights may be used by new actors to gain access to water. The water rights system as administered by the Tanzanian government in the Pangani basin provides the legal means for powerful actors to dispossess existing users. As demonstrated in Chapter 5, powerful cities in the Pangani basin selectively used the law to gain leverage over water control. In other cases the water right systems has led to struggles over legitimacy (Chapter 8). The legitimacy of the state-based water rights system apparently is questioned by several actors. The case of Nduruma catchment (Chapter 8) presents a situation where small scale users appeal to customary principles while large-scale irrigators attempt to gain water access using the state's statutory water law. Although the estates have location advantage, their 'official water right' does not go unchallenged by the downstream smallholder farmers. These farmers demand that allocation should be rotational and take into account supply variability and not the absolute values specified in the government water right. During the dry season or in times of low flows the estates water allocations are often reduced from 24 hours (as stated in the state-issued permit) to 5-7 hours per day. The estates are increasingly being pushed out of surface water with many of them switching to groundwater use. Groundwater use, availability as well as its interaction with surface water in the catchment is not yet well understood. The local resource users do not yet see the use of groundwater by large commercial estates as a threat to surface water and groundwater availability.

There seems to be a problem with enabling meaningful participation by the resource users in decision making related to catchment water management. The operationalisation of the catchment forum concept as demonstrated in Chapter 4 faced significant challenges. In the Kikuletwa catchment it proved difficult to define the most appropriate hydrological management unit for decision-making that was able to fit well with the political-administrative territories. The basin water board and development partners tried to resolve the problem of administrative boundaries and institutional fit by selecting users' representatives at the river level from each of the administrative wards comprising a particular river. But this was insufficient to integrate customary arrangements into the state-led governance structure. Institutional arrangements created in water scarce zones of a catchment, e.g. in the lowlands (Chapter 3), succeeded in minimising water conflicts between farmers there. Water dialogue arrangements created at larger spatial scales, as demonstrated in Hingilili (Chapter 3) and Kikuletwa (Chapter 4), have so far failed to function properly. The ready explanation is that the latter organisations were not properly linked to existing institutional arrangements. This is likely to be correlated with the fact that there is no need nor incentive for upstream furrows to engage with their downstream counterparts, because of their location advantage. The messy overlapping jurisdictions between state-led and locally created institutions in the catchment mean the new structures could only be layered on top of pre-existing local institutional arrangements. The inability of an apex catchment organisation to function as designed (Chapter 3) highlights ambiguities of institutional overlaps and linkages between local and state forms at a sub-catchment level, echoing Metha et al (1999)'s 'messy middle'. The Kikuletwa case showed how new institutional arrangements are highly contested.

This research showed that the concept of institutional nesting does not necessarily work at catchment level. To overcome the large spatial extent of catchments, modularisation into smaller sub-units has been proposed, which would allow the creation of polycentric governance that nests local arrangements. In the Kikuletwa catchment (Chapter 4), the water users do not see how the newly introduced sub-catchment WUA is linked to their own governance arrangements (e.g. furrow and river committees) and constantly ask "how do we benefit from paying memberships and annual fees to the WUA?" An important question here is whether local solidarity based institutional arrangements can really be upscaled while at the same time downscaling state-led interventions. Institutional nesting appears to be the most logical approach but as demonstrated in Chapter 4, it still raises the question of institutional fit. Moreover, any new arrangement is still likely to undergo processes of institutional bricolage producing unpredictable and sometimes negative impacts on water access (see Chapter 3). Besides, local self governing arrangements as demonstrated in Chapter 3 and 6 are not necessarily fair. The general conclusion is that resolving the problem of institutional fit while integrating customary arrangements with the state-led governance structure requires careful analysis of local structures, and a good understanding of their strengths and limitations. As the local structures may lie outside the water realm, this also implies that those who design interventions for the water sector should be able and willing to look beyond water.

The second sub-objective: attempt to understand local water management practices: why they emerge, and how they function and are being sustained. In other words attempt to understand the mechanisms that drive cooperation at the local level (e.g. turn taking in villages – between farmers sharing a furrow, between two furrows, between neighbouring villages, between distant villages and within a catchment).

This thesis also investigated the potential of upscaling local water management practices. The emergence of self governing arrangements was found to be driven by among others: environmental shocks (e.g. droughts), gradual population growth and economic development. Local level innovation with institutional arrangements for water sharing often emerged around the creation of hydraulic property and/or was negotiated to secure more water flow for downstream users. Through the process of institutional bricolage (Cleaver 2002), it has been demonstrated that these arrangements often build on and borrow from existing institutions within society, which are frequently unrelated to water. The hydraulic position of the various actors in a catchment (upstream or downstream) adds a complicating dimension to these institutional dynamics and may be considered as a driver for institutional innovation. In the cases studied it was always the downstream users that initiated the process of institutional change in a catchment. It may be hypothesised that institutional innovation in a catchment gradually crawls upstream as the need for coordination grows.

In Chapter 6, it was demonstrated that despite a relatively high level of inequality of access to land (Gini coefficient of 0.58), irrigation canals in Makanya catchment are still being sustained. It is argued that this inequality might be understood as the outcome of underlying dynamics that limit excesses: if land and water become concentrated in too few hands, irrigators may opt out and shift their efforts to another canal, making the mobilization of labour for maintenance more problematic, and hence possibly leading to system collapse.

But this inequality alone appears insufficient to explain how these canal organisations have been able to endure. Heterogeneity is considered to be an additional factor that has led to mutual dependencies among the irrigators. The findings are relevant to our understanding of the functioning of water institutions more generally in three ways. First, inequality among users is not naturally a deterrent to collective action, and may in specific cases even be considered a resource that can help initiate and maintain the collective good. Second, mutual dependencies that exist between users and user groups in a water system may similarly be considered to be a vital resource, e.g. farmers owning plots in several furrows systems. Making such dependencies explicit or even consciously increasing them may help to stimulate collective action. Third, the combination of inequality and interdependencies may give rise to emerging dynamics that can explain sustained collective action in situations of water asymmetry. This is exemplified by the fact that the canal organisations studied in this thesis have been functioning for several generations, and have not collapsed. This implies that some of the keys to enduring water institutions may lie outside the water sector.

The above findings are confined to relatively small spatial and social scales, involving irrigators from one village. In such situations there may be inhibitions to unilateral action due to social and peer pressure. Proximity may thus be a necessary condition for collective action in water asymmetrical situations to emerge. At larger spatial scales and over greater distances, for example when considering entire catchment areas or river basins, this is likely to be different. The social relationships that could promote collective action in such larger spatial scales have hardly been studied but could include social relations such as inter-village marriages, church groups, seasonal or longer-term migration patterns, pastoral movements, regional markets as well as formal representation in local and district and higher-level government. A better understanding of such relationships, their inequalities and heterogeneity, may help identify incentives for collective action when the need arises. Those characteristics could well be powerful complementary arrangements to formal, top-down established basin organizations.

The emergence of river committees (RCs) to manage water allocation between users along a river may be considered an attempt by the local resource users to overcome the effect of water asymmetry and inequality at larger spatial scales. The thesis described and analysed the emergence and functioning of RCs in several sub-catchments of the Pangani river basin (Chapter 7). The RCs emerged in response to competition and conflicts over limited water induced by, among others, increased frequency of low flows, natural population growth, markets for agricultural produce, and government policies. Their functioning was strongly influenced by: (1) the number of users sharing a tributary; (2) differences between the users in terms of type of water use (irrigation, livestock, domestic) and size of water use (commercial vs. subsistence); (3) socio-cultural differences between the users, e.g. different ethnic groups; (4) location (hydraulic) advantage and spatial distance between the users; (5) the straddling of administrative boundaries; and (6) the presence of markets for (high-value) agricultural products. It was observed that the entrepreneurial farming opportunities in Arusha municipality promote a new kind of water rationality. Whereas in former times water users seem to have defined their self-interest in terms of broader social, spatio-temporal interdependencies (Van der Zaag, 2007, after Alam, 1998), this is now changing. Upstream users now view water as a source of private wealth, rather than a resource that requires collective action to generate and maintain a stream of benefits.

Third, develop a game theoretic model for considering alternative scenarios for collective catchment management. This thesis also explored the emergence and functioning of local water management practices using game theoretic consideration. Following detailed study of cases in the Pangani river basin (Chapters 6-8) a first attempt was made to develop a game-theoretic model of irrigation canal cleaning (Chapter 9). The canal cleaning game was developed to improve our understanding of how local arrangements can emerge and endure, under condition of inequality in landholding, location asymmetry, differences in the costs and benefits of, and mutual dependence on collective action. The simple game provides some general and preliminary insights into the functioning of self-governing irrigation systems. This will be further explored in a follow up research.

In light of these findings I can now address the main objective of this thesis: *to explore conditions for reconciling state-led institutional arrangements and local water management practices.* Drawing from the findings of this thesis, it may be argued that local solidarity approaches function best at the scale in which they are currently found. Beyond the small spatial scale, however, they may be difficult to initiate and sustain and are likely to collapse. I can also conclude from this research that an intervention that attempted to establish a hierarchical structure that nests local water management arrangements was not successful in the catchment studied. However there is a possibility to integrate state-led river basin management structure with local water management arrangements. In the Pangani basin, sub-catchment water users associations do not lead to effective water management as was shown in Chapter 4. I also find that the river committee is the most promising locally evolved institution that can link state-led and locally created water institutions. The river committees would continue to manage water allocation between users within their river reach and could be issued a collective water use right on the condition to ensure some minimum outflow during the dry season for downstream use. The Basin Water Board could then concentrate its efforts on monitoring the outflow from each river, and penalise committees if the minimum flow conditions were violated. This way the local water allocation systems could complement the state regulatory water rights.

10.2 Contribution to Theories, Concepts and Methodology

This thesis used several concepts and theories to study conditions for reconciling state-led institutional arrangements and local water management practices. In so doing I contributed to existing theories and concepts related to catchment management. Theories and concepts such as institutional bricolage (Cleaver 2002), echelons of water rights analysis (Boelens 2008), water property rights (Bromley 1992), and hydraulic property creation (Coward 1986a) were mainly used to analyse the selected case studies. However, we did highlight their limitations. The following theories and concepts were modified and will be discussed in this section: 1) river basin development trajectory (Molle 2003); 2) institutional design principles (Ostrom 1993); and 3) collective action over common pool resources (Olson 1965, Baland and Platteau 1999). I will also make critical observations on: polycentric governance; catchment/multi-stakeholder forums; the implications of establishing state-led water rights in the Pangani basin; and the application of game theory to understand water institutions.

The development path of a river basin over time and space was first conceptualised by Keller et al. (1998) as a linear path and sequence of actors' responses. This was later modified by Molle (2003) in a graphical representation that acknowledges the variety of micro/local and macro/global responses to water problem. In this thesis we expanded Molle's (2003) typology of basin actors' responses by explicitly introducing a meso layer which depicts the interface where State-level and local-level initiatives and responses are played out; and we focused on how this interaction finds expression in the creation and modification of hydraulic property rights. To capture the link between property right and the different responses at the micro- and macro-level, we added property relation to Molle's framework, intersecting these two levels. The expanded typology of basin actors' responses provided insight into the dynamic situation in the Pangani river basin.

The thesis showed that not all the eight design principles proposed by Ostrom (1993) are necessary for a water institution to be effective and to endure over time. Unlike Ostrom's claim that well-defined boundaries are needed, we find boundary definitions as used by the resource users context-dependent, ambiguous and fluid; they are thus not clearly defined. By not entirely closing the resource boundary, the users lower the transaction cost of controlling the boundaries and also allow future demands to be met in the face of increasing resource variability. The arrangements do not fully comply with the principle that all affected must take part in rule creation and modification. Also, members may consciously abscond from meetings to avoid ratifying binding agreements. Monitoring is at best ad hoc, and downstream users have to invest extra resources to secure allocation. The difference in the performance of local water institutional arrangements could not be explained by referring to Ostrom's eight design principles alone.

This thesis clarified how collective action emerges and is sustained in settings where water asymmetries exist. We provided conceptual clarity to the uses of the terms heterogeneity and inequality in relation to collective action over common pool resources. In many studies heterogeneity is conflated with inequality (Poteete and Ostrom 2004). I showed that distinguishing between inequality and heterogeneity provides conceptual clarity and yields a better understanding of their linkages. Unlike most research on collective action in which water asymmetry, inequality and heterogeneity are seen as risks to collective action, this thesis showed that they dynamically interact and may give rise to interdependencies between water users which may facilitate coordination and collective action.

The thesis further showed that although the catchment forum concept may be considered a good idea for effective participatory management of water stressed basins it is difficult to implement in an African catchment like the Pangani basin. Challenges identified in the thesis include large and complex catchment characteristics, both in term of coverage and hydrology; diverse stakeholders; and messy institutional arrangements with overlapping jurisdiction between state-led and locally created institutions.

The relevance of the polycentric water governance approach as a framework for integrating local and state-led institutions was also studied (Andersson and Ostrom 2008, Lankford and Hepworth 2010). Modularisation of the large catchment into sub-catchments to allow polycentric governance that nests local arrangements did not work in the Kikuletwa catchment. The sub-catchment water users association formed by the Pangani Basin Water Board and partners only created more governance layers without effectively integrating locally evolved arrangements such as the river committees. The resource users did not see how the newly created forums were linked to their own governance arrangements. Institutional nesting or polycentric governance is therefore of no relevance for African rivers if not linked to pre-existing local arrangements, be they formal or informal. In the Pangani basin there may not have been a need to create new governance forums, exisiting river committees are in fact dialogue forums and should be seen as such. All the river committees found active in the Pangani basin include large-scale and small-scale irrigators, wards and village administrations, upstream and downstream users relying on the same river as members. River committees manage the allocation of water between the users of a part of a river and appear to be able to solve the coordination challenge experienced by upstream and downstream, and large-scale and small-scale farmers. Their success lies in the fact that they are considered legitimate by the water users and the local government institutions. The institutional form of a River Committee thus bridges local rules and statute law with respect to water.

This thesis also discussed in detail the historical development of state-led water rights in the Pangani river basin. Thus the thesis contributes to the global discourses on efficiency, IWRM, and water rights, as used to justify state intervention in the water sector. The water right system as implemented in the Pangani basin is difficult to enforce and control, and so far has not led to efficient water use. About 3400 users have been identified by the Pangani basin water board but over 50% (1800) are without formal water permits. This is still a tiny minority of all the water users in the basin. The Pangani Basin Water Board has no capacity to monitor water use in the catchment, and also cannot ensure that the numerous users (small and large) stick to the official water allocation. Even if only large users were to be issued water use permits (Maganga et al. 2004), flow variability caused by unpredictable rainfall, and recurrent droughts still make the fixed volumetric water use rights ineffective. The thesis offers complementary solutions to the water allocation challenges. Instead of fixed volumetric rights we propose a proportional allocation system as an alternative by which limited water resources can be fairly allocated, e.g. permitted abstractions are reduced in proportion to the expected shortfall in river flow. The exact amounts (quantity or duration of use) by which individual allocations are reduced would be negotiated by the users at the river level. This is already being practiced by the water users in Nduruma sub-catchment (Chapter 8). Through negotiated water allocation, managed at the river level, e.g. by river committees, some of the problems can be solved. This is not to say that local allocation systems should replace the state water right system; they should complement the state regulatory effort.

The application of participatory gaming to understand water conflicts and cooperation was shown to be an effective research tool. A preliminary attempt was made in this thesis to use a game theoretic approach to understand the development of local water institutions around an irrigation canal. At the limited scale at which it was applied, the canal cleaning game proved useful and will be extended in a follow up research combining game theory and agent-based modelling.

To achieve the objectives of this thesis, the research involved in-depth descriptions and analyses of selected cases within the Pangani river basin. The strategy used was inspired by the 'follow the water' approach. By employing a mixture of techniques to achieve the research objectives I was able to capture the dynamic interactions at the local level and produce narratives. Role play games administered through feedback workshops allowed me to engage in multiple dialogues with the object of research, and all this based on a meticulous cartography of irrigation canals and irrigated plots and zones. The innovative part of the research methodology used is best summed up by the metaphor of a big house with many semi-detached rooms, each of which can function independently but combine to form the complex whole. Using this approach, I was able to study several cases, moving from a detailed case study of one irrigation canal to a large catchment like the Kikuletwa comprising a complex system of many tributaries with many diversion canals. The approach allowed me to slide back and forth between the process of state-led intervention and local water management practices. This way I was able to engage with and contribute to the projects of several actors. During the course of my field work, I was invited by the Pangani Basin Water Board and partner NGOs to make presentations on research findings, observe, and to actively facilitate forum workshops. I have engaged with the resource users in the catchment, and continue to interact with leaders of river committees in the catchment. The close collaboration I developed with the Pangani Basin Water Board/Office, development NGOs active in the Pangani Basin, and the water users, proved useful in communicating the research findings.

This thesis raised a number of issues that require further research and analysis. First village governments do play important roles in water management and are actively involved in the activities of the existing furrow committees, and river committees. Village governments are often called upon to solve water conflicts that could not be resolved by furrow committees. The villages are nested government structures with legitimacy at the local level. However the process of creating sub-catchment water users associations stopped at the ward level. Research is needed to understand the role village government can play in addressing competition over water at larger spatial scales. Second, the Pangani basin is a closing basin with many of its tributaries now only flowing for part of the year. Investment in new hydraulic infrastructures such as storage dams is one route for resolving water competition but with potentially large downstream impacts; this needs to be investigated. In other basins in Sub-Saharan Africa there may still be room for infrastructure development (e.g. building new dams and pumping more groundwater) but for closing basins like the Pangani, such efforts are likely to speed up the closing process. In closing basins, interventions aimed at capturing more water should be done following a thorough understanding of the basin's hydrological and ecological interconnections (Molle et al. 2010). Third, this thesis did not discuss in-depth the dynamic of gender, inequality and access to water. Leadership of local as well as state-led water management organisations in the Pangani basin are male dominated and in such a situation equity and fairness with respect to gender may be compromised. Better reconciliation of state-led and local water management arrangements with fewer opportunities and better checks for the more powerful to widen inequities may as well benefit women and other marginalized groups; this requires further research. Finally to provide more insight into the functioning of self-governing institutions, further research is needed: 1) to describe phenomena of water asymmetry, inequality and heterogeneity at larger spatial scales, and to analyse under which circumstances they occur; and 2) to verify the relation between inequality of access to land and water in furrow systems and the collective ability to share water and mobilize labour for maintenance at many other furrow systems in order to generalize the findings of this thesis, not only in the Pangani but also in the Rufiji river basin in Tanzania, as well as in other African countries, such as Kenya and Mozambique, and perhaps even in other continents, such as in Nepal.

10.3 CRITICAL REFLECTION ON STRENGTH AND LIMITATION OF THE RESEARCH

Coming to the end of this thesis it is important to reflect on the strength and limitations of the research. In this section I will discuss: 1) my role as an outsider (foreigner); 2) action research; 3) my background as a civil engineer; 4) gender; and 5) the potential of agent-based modelling and participatory gaming.

Coming from Uganda, I went to Tanzania having little or no knowledge of Swahili (the official language of the country). As it is true for any foreign researcher, this problem can be solved with the help of a native speaker. For me this was initially not easy. In Makanya catchment where I started my field research it was not easy to find someone who could effectively translate English to Swahili and vice versa. At times I had to rely on two translators, one for English to Swahili and the other for Swahili to English. More importantly, using a translator is like putting a windscreen between you and the interviewee. The translator first filters the information and then provides what he/she perceives is relevant for you and the interviewee. Sometimes important pieces of information may be lost. I used triangulation (field observation, mapping and extended discussions) to overcome this problem. However, I found learning the language was particularly helpful as this allowed me to combine my firsthand understanding of the respondents viewpoints with what was being translated and seek further clarification if necessary. The fact that I am from Uganda ("foreigner") did not hinder my field activities, perhaps because of the historical connection between the two countries or just because I am an African. Tanzanians are generally very welcoming - they allowed me to enter their life worlds. Attending water meetings, actively participating in activities and assisting farmers where I could, such as providing transport to farmers going to negotiate water allocation arrangements with upstream villages, proved useful in gaining the trust of the communities I researched.

Making my research work of relevance to the people I worked with was particularly useful in obtaining additional information. At the furrow level I provided irrigation land maps to farmers and the village office, while at the river level I conducted several feedback sessions with farmers. I was able to gain trust by providing transport to furrow leaders and village elders to attend water negotiation in upstream villages. I attended several furrow water allocation meetings. At the basin level I supported through workshop facilitations, presentationg of research findings at partners meetings and sharing information (e.g. maps, water users' details etc) I have collected with the Pangani Basin Water Office. The Basin Water Officer was particularly welcoming, he allowed me access to information I wouldn't otherwise have accessed.

The fact that I am a civil engineering by training and not a social scientist certainly has influenced the outcome of the research. There could be details that I overlooked or I could have used a different approach to engage with the actors during the research. However, my technical background and curiosity to learn new ways of observing and collecting information proved useful. I was able to combine both technical skills and social methods to gain in-depth knowledge of the biophysical, material and social connections of the water systems I studied.

Gender in water resources management was a limitation in this research. First, I am male. This may have biased me to see things in certain ways. Second, leadership positions in local water management organisations are male dominated in the study area, meaning most of my respondents were male. However, during the research, I purposively selected female members of the furrow organisations for interviews and discussions.

Finally, understanding water institutions that have evolved over time and space requires close observations. This thesis provided insight into the emergence and functioning of such locally evolved arrangements. However, more data is still required on the functioning of these systems at larger spatial scales. A preliminary attempt was made using game theoretic approach to understand how local arrangements arose and could endure, given the inequality in landholding, location asymmetry, and differences in the costs and benefits of collective action. Although this proved useful, there is a need to capture the dynamics in the system. It is hypothesized that this is possible considering agent-based modelling with participatory gaming with real farmers. The initial plan was to complement the findings from the detailed case studies with agent based modelling. The institutional arrangements for sharing water resources were to be constructed into an agent-based model informed by field data and discussions with communities. This could have allowed different scenarios and new sets of questions to be developed to improve further our understanding of local water management practices. As mentioned in the introductory chapter of the thesis this approach is still valid to me. A recursive process that combines modelling and in-depth case studies is very challenging and requires a longer time frame. The research strategy I finally adopted proved useful in understanding the dynamics of state-led and local water management institutions.

Overall, the Pangani river basin provided a unique opportunity to study the emergence and evolution of endogenous water governance institutions. The basin proved a suitable living laboratory for studying local level institutional arrangements and state intervention.

REFERENCES

Adams, W. M., Potkanski, T. and Sutton, J. E. G. 1994. Indigenous farmer-managed irrigation in Sonjo, Tanzania. *The Geographical Journal,* 160(1), 17 - 32.

Adams, W. M., Watson, E. E. and Mutiso, S. K. 1997. Water, rules and gender: water rights in an indigenous irrigation system, Marakwet, Kenya. *Development and Change,* 28(4), 707 - 730.

Adhikari, B. and Lovett, J. C. 2006. Institutions and collective action: does heterogeneity matter in community-based resource management? *The Journal of Development Studies,* 42(3), 426-445.

Agrawal, A. 2001. Common property institutions and sustainable governance of resources. *World Development,* 29(10), 1649-1672.

Andersson, K. and Agrawal, A. 2011. Inequalities, institutions, and forest commons. *Global Environmental Change,* 21(3), 866-875.

Andersson, K. P. and Ostrom, E. 2008. Analyzing decentralized resource regimes from a polycentric perspective. *Policy sciences,* 41(1), 71-93.

Andersson, R., Wanseth, F., Cuellar, M. and Von Mitzlaff, U., 2006. *Pangani Falls redevelopment project in Tanzania.* Stockholm: Swedish International Development Cooperation Agency (SIDA).

Araral Jr., E. 2009. What explains collective action in the commons? Theory and evidence from the Philippines. *World Development,* 37(3), 687-697.

Asmal, K., 2003. Arid African upstream safari: A transboundary expedition to seek and share new sources of water. *In:* Dooge, J., Delli Priscoli, J. and Llmas, M. R. eds. *Water and Ethics.* Paris: UNESCO.

Baffes, J., 2003. *Tanzania's coffee sector: constraints and challenges in a global environment.* Washington DC: World Bank.

Baland, J. M. and Platteau, J. P., 1996. *Halting degradation of natural resources: Is there role for rural communities?* Oxford: FAO/Clarendon Press.

Baland, J. M. and Platteau, J. P. 1999. The ambiguous impact of inequality on local resource management. *World Development,* 27(5), 773-788.

Bardhan, P. and Dayton-Johnson, J., 2002. Unequal irrigators: Heterogeneity and commons management in large-scale multivariate research. *In:* Ostrom, E. ed. *The drama of the commons.* Washington, DC: National Academy Press.

Berndes, G. 2002. Bioenergy and water—the implications of large-scale bioenergy production for water use and supply. *Global Environmental Change,* 12(4), 253-271.

Bhattarai, M., Pant, D. and Molden, D. 2005. Socio-economics and hydrological impacts of melamchi intersectoral and interbasin water transfer project, Nepal. *Water Policy*, 7(2), 163-180.

Blomquist, W., 2009. Multi-level governance and natural resource management: The challenges of complexity, diversity, and uncertainty. *In:* Beckmann, V. and Padmanabhan, M. eds. *Institutional sustainability: Political economy of agriculture and the environment - Essays in honour of Konrad Hagedorn.* Springer.

Bodansky, D. M. 1999. The legitimacy of international governance: a coming challenge for international environmental law? *The American Journal of International Law*, 93(3), 596-624.

Bodansky, D. M., 2007. The concept of legitimacy in international law. *Research paper series.* University of Georgia School of Law, 1-8.

Boelens, R. 2008. Water rights arenas in the Andes: upscaling networks to strengthen local water control. *Water Alternatives*, 1(1), 48-65.

Boelens, R. and Davila, G., eds., 1998. *Searching for equity: Conceptions of justice and equity in peasant irrigation.* Assen: Van Gorcum.

Boettke, P. J., Coyne, C. J. and Leeson, P. T. 2008. Institutional Stickiness and the New Development Economics. *American Journal of Economics and Sociology*, 67(2), 331-358.

Bohensky, E. and Lynam, T. 2005. Evaluating responses in complex adaptive systems: Insights on water management from the Southern African Millennium Ecosystem Assessment (SAFMA). *Ecology and Society*, 10(1), 11.

Bolding, A., 2004. *In hot water: A study on sociotechnical intervention models and practices of water use in smallholder agriculture, Nyanyadzi catchment, Zimbabwe.* (PhD thesis). Wageningen University.

Bossio, D., Jewitt, G. and van der Zaag, P. 2011. Smallholder system innovation for integrated watershed management in Sub-Saharan Africa. *Agricultural Water Management*, 98(11), 1683-1686.

Boyd, R., Gintis, H. and Bowles, S. 2010. Coordinated Punishment of Defectors Sustains Cooperation and Can Proliferate When Rare. *Science*, 328(5978), 617-620.

Brody, S. D., 2008. *Ecosystem planning in Florida: Solving regional problems through local decision making.* Aldershot, UK: Ashgate Press.

Bromley, D. W. 1989. Property relations and economic development: the other land reform. *World Development*, 17(6), 867 - 877.

Bromley, D. W. 1992. The commons, common property and environmental policy. *Environmental and Resource Economics*, 2(1), 1 - 17.

Bromley, D. W. 1997. Constitutional political economy: property claims in a dynamic world. *Contemporary Economic Policy*, 15(4), 43–54.

Bruns, B. 2007. Irrigation water rights: options for pro-poor reform. *Irrigation and drainage*, 56(2-3), 237-246.

Bruns, B. R., 2009. Metaphors and methods for institutional synthesis. *Workshop in Political Theory and Policy Analysis: Panel on Water Resource Governance and Design Principles.* Indiana University, Bloomington.

Bruns, B. R. and Meinzen-Dick, R. S. 2001. Water rights and legal pluarlism: four contexts for negotiation. *Natural Resources Forum*, 25(1), -10.

Buchanan, J. M. 1965. An Economic Theory of Clubs. *Economica*, 32(125), 1-14.

Budescu, D. V. and Au, W. T. 2002. A model of sequential effects in common pool resource dilemmas. *Journal of Behavioral Decision Making,* 15(1), 37–63.

Burawoy, M. 1998. The extended case method. *Sociological Theory,* 16(1), 4-33.

Burra, R. and Van den Heuvel, K., 1987. Traditional irrigation in Tanzania: a historical analysis of irrigation tradition and government intervention.

Cardenas, J. C., Rodriguez, L. A. and Johnson, N. 2011. Collective action for watershed management: field experiments in Colombia and Kenya. *Environment and Development Economics,* 16(Special Issue 03), 275-303.

Carlsson, E., 2003. *To have and to hold: Continuity and change in property rights institutions governing water resources among the Meru of Tanzania and the Bakgatla in Botswana; 1925 - 2000.* (PhD thesis). Lund University.

Celio, M., Scott, A. C. and Giordano, M. 2010. Urban-agricultural water appropriation: the Hyderabad, India case. *The Geographical Journal,* 176(1), 39-57.

Challen, R., 2000. *Institutions, transaction costs and environmental policy: Institutional reform for water resources.* Cheltenham, UK: Edward Elgar.

Characklis, G. W., Kirsch, B. R., Ramsey, J. and Dillard, K. E. M. 2006. Developing portfolios of water supply transfers. *Water Resources Research,* 42(W05403).

Cleaver, F. 1999. Paradoxes of participation: questioning participatory approaches to development. *Journal of International Development,* 11, 597 - 612.

Cleaver, F. 2000. Moral ecological rationality, institutions and the management of common property resources. *Development and Change,* 31(2), 361-383.

Cleaver, F. 2002. Reinventing institutions: Bricolage and the social embeddedness of natural resource management. *The European Journal of Development Research,* 14(2), 11-30.

Cleaver, F. and Franks, T., 2005. How institutions elude design: River basin management and sustainable livelihoods. University of Bradford Centre for International Development, UK, 21.

Cleaver, F. and Toner, A. 2006. The evolution of community water governance in Uchira, Tanzania: The implications for equality of access, sustainability and effectiveness. *Natural Resources Forum,* 30(3), 207-218.

Coward, E. W. J., 1986a. Direct or indirect alternatives for irrigation investment and the creation of property. *In:* Easter, K. W. ed. *In irrigation investment, technology and management strategies for development, ed.*: Boulder, Westview Press, pp. 491 - 508.

Coward, E. W. J., 1986b. State and locality in Asian irrigation development: the property factor, ed. *In:* Nobe, K. C. and Sampath, R. K. eds. *Irrigation management in developing countries : current issues and approaches.* Boulder, Westview Press, pp. 225 - 244.

de Fraiture, C., Molden, D. and Wichelns, D. 2010. Investing in water for food, ecosystems, and livelihoods: An overview of the comprehensive assessment of water management in agriculture. *Agricultural Water Management,* 97(4), 495-501.

de Fraiture, C. and Wichelns, D. 2010. Satisfying future water demands for agriculture. *Agricultural Water Management,* 97(4), 502-511.

Dietz, T., Ostrom, E. and Stern, P. C. 2003. The struggle to govern the commons. *Science,* 302(5652), 1907-1912.

Dombrowsky, I. 2010. The role of intra-water sector issue linkage in the resolution of transboundary water conflicts. *Water International Journal of River Basin Management,* 35(2), 132–149.

Douglas, M., 1986. *How institutions think.* Syracuse: Syracuse University Press.

Dube, D. and Swatuk, L. A. 2002. Stakeholder participation in the new water management approach: a case study of the Save catchment, Zimbabwe. *Physics and Chemistry of the Earth, Parts A/B/C,* 27(11-22), 867-874.

Ekstrom, J. A. and Young, O. R. 2009. Evaluating functional fit between a set of institutions and an ecosystem. *Ecology and Society,* 14(2), 16.

Enfors, E., 2009. *Traps and transformations: Exploring the potential of water system innovations in dryland sub-Saharan Africa.* (PhD thesis). Stockholm University.

Enfors, E. and Gordon, L. J. 2007. Analysing resilience in dryland agro-ecosystems: a case study of the Makanya catchment in Tanzania over the past 50 years. *Land Degradation and Development,* 18(6), 1 - 16.

Ensminger, J. 1990. Co-oping the elders: the political economy of state incorporation in Africa. *American Anthropological Association,* 92(3), 662 - 675.

Epstein, J. M., 2006. *Generative social science: studies in agent-based computational modelling.* Princeton: Princeton University Press.

EWURA, 2010. *Water utilities performance report for the year 2008/2009 part A: Urban Water Supply and Sewerage Authories (UWSA's) in Tanzania*

Falkenmark, M. 2007. Shift in thinking to address the 21st century hunger gap. *Water Resources Management,* 21(1), 3-18.

Falkenmark, M. and Folke, C. 2002. The ethics of socio-ecohydrological catchment management: towards hydrosolidarity. *Hydrology and Earth System Sciences,* 6(1), 1-9.

Falkenmark, M. and Molden, D. 2008. Wake up to realities of river basin closure. *Water Resource Management,* 24(2), 201 - 215.

Faurès, J.-M. and Santini, G., eds., 2008. *Water and the rural poor: interventions for improving livelihoods in Sub-Saharan Africa.* Rome: FAO.

Faysse, N. 2006. Troubles on the way: An analysis of the challenges faced by multi-stakeholder platforms. *Natural Resources Forum,* 30(3), 219-229.

Fischhendler, I., Feitelson, E. and Eaton, D. 2004. The short-term and long-term ramifications of linkages involving natural resources: the US - Mexico transboundary water case. *Environment and Planning C: Government and Policy,* 22(5), 633-650.

Fivas, 1996. *Power conflicts: Norwegian hydropower developers in the Third Worlds* [online]. Fivas. Available from: http://www.fivas.org/sider/tekst.asp?side=121 [Accessed July 24 2008].

Fleuret, P. 1985. The social organisation of water control in the Taita hills, Kenya. *American Anthropological Association,* 93(1), 91-114.

Galvan, D. 1997. Institutional syncretism and the articulation of modes of production in rural Senegalese land tenure relations. *African Rural and Urban studies,* 4(2 -3), 59-98.

Galvan, D., 2007. Syncretism and local-level democrarcy in Rural Senegal. *In:* Galvan, D. and Sil, R. eds. *Reconfiguring institutions across time and space: Syncretic responses to challenges of political and economic transformation.* Palgrave Macmillan.

Gearey, M. and Jeffrey, P. 2006. Concepts of legitimacy within the context of adaptive water management strategies. *Ecological Economics,* 60(1), 129-137.

Gibson, C. C., Ostrom, E. and Ahn, T. K. 2000. The concept of scale and the human dimensions of global change: a survey. *Ecological Economics,* 32(2), 217 - 239.

Gillingham, M. E. 1999. Gaining access to water: Formal and working rules of indigenous irrigation management on Mount Kilimanjaro, Tanzania. *Natural Resources Journal,* 39(3), 419 - 441.

Gray, R. F., 1963. *The Sonjo of Tanganyika: An anthropological study of an irrigation-based society.* London: Oxford University Press.

Grove, A. 1993. Water use by the Chagga on Kilimanjaro. *African Affairs,* 92(368), 431-448.

Gupta, J. and Van der Zaag, P. 2008. Interbasin water transfers and integrated water resources management: where engineering, sciences and plitics interlock. *Physics and Chemistry of the Earth,* 33(1), 28-40.

GWP, 2000. *Integrated water resources management.* Stockholm, Sweden: GWP.

Hackett, S., Schlager, E. and Walker, J. 1994. The Role of Communication in Resolving Commons Dilemmas: Experimental Evidence with Heterogeneous Appropriators. *Journal of Environmental Economics and Management,* 27(2), 99-126.

Håkansson, T., N. 1998. Rulers and rainmakers in precolonial South Pare, Tanzania: Exchange and ritual experts in political centralization. *Ethnology,* 37(3), 263 - 283.

Hearne, R. R. 2007. Water markets as a mechanism for intersectoral water transfers: the Elqui Basin in Chile. *Paddy and Water Environment,* 5(4), 223-227.

Hirschman, A., O, 1970. *Exit, voice, and loyalty: responses to decline in firms, organisations, and states.* Cambridge, MA: Harvard University Press.

Horst, L., 1998. *The dilemmas of water division: considerations and criteria for irrigation system design.* Colombo, Sri Lanka: International Water Management Institute.

Howitt, R. E., 1998. Spot prices, option prices and water markets: analysis of emerging markets in California. *In:* Easter, K. W., Rosegrant, M. W. and Dinar, A. eds. *Markets for water: potential and performance.* Norwell, Massachusetts: Kluwer Academic.

IUCN Eastern and Southern Africa Programme, 2009. *The Pangani river basin: A situation analysis 2nd Edition.*

Janssen, M., Anderies, J. and Joshi, S. 2011a. Coordination and cooperation in asymmetric commons dilemmas. *Experimental Economics,* 14(4), 547-566.

Janssen, M. A., Anderies, J. M. and Cardenas, J.-C. 2011b. Head-enders as stationary bandits in asymmetric commons: Comparing irrigation experiments in the laboratory and the field. *Ecological Economics,* 70(9), 1590-1598.

Janssen, M. A., Holahan, R., Lee, A. and Ostrom, E. 2010. Lab Experiments for the Study of Social-Ecological Systems. *Science,* 328(5978), 613-617.

Jaspers, F. G. W. 2003. Institutional arrangements for integrated river basin management. *Water Policy,* 5(1), 77-90.

JICA, 1984. *Feasibility study on the Mkomazi valley area irrigation development project.*

Jones, E. C. 2004. Wealth-based trust and the development of collective action. *World Development,* 32(4), 691-711.

Kashaigili, J. J., Kadigi, R. M. J., Sokile, C. S. and Mahoo, H. F. 2003. Constraints and potential for efficient inter-sectoral water allocations in Tanzania. *Physics and Chemistry of the Earth,* 28(20-27), 839 - 851.

Keller, J., Keller, A. and Davids, G. 1998. River basin development phase and implications of closure. *Journal of Applied Irrigation Science,* 33(2), 145 - 163.

Kemerink, J. S., Ahlers, R. and Van der Zaag, P. 2009. Assessment of the potential for hydro-solidarity within plural legal conditions of traditional irrigation systems in northern Tanzania *Physics and Chemistry of the Earth*, 34(13-16), 881-889.

Kikula, I. S., 1997. *Policy implications on environment: the case of villagisation in Tanzania.* Nordic Africa Institute.

Kissawike, K., 2008. *Irrigation-based livelihood challenges and opportunities: A gendered technology of irrigation development intervention in the Lower Moshi irrigation scheme in Tanzania.* PhD Thesis (PhD). Wageningen University.

Komakech, H. C., Condon, M. and Van der Zaag, P. 2012a. The role of statutory and local rules in allocating water between large- and small-scale irrigators in an African river catchment. *Water SA*, 38(1), 115-125.

Komakech, H. C., Miller, J. H. and Van der Zaag, P., 2012b. A game theoretic analysis of evolution of cooperation in small-scale irrigation canal system. *Paper prepared for the International Conference on Fresh Water Governance for Sustainable Development.* Drakensburg Sports Resort, South Africa.

Komakech, H. C., Mul, M. L., Rwehumbiza, F. and Van der Zaag, P. 2011a. Water allocation and management in an emerging spate irrigation system in Makanya catchment, Tanzania. *Agricultural Water Management*, 98(11), 1719-1726.

Komakech, H. C. and Van der Zaag, P. 2011. Understanding the emergence and functioning of river committees in a catchment of the Pangani basin, Tanzania. *Water Alternatives*, 4(2), 197-222.

Komakech, H. C. and Van der Zaag, P. 2013. Polycentrism and pitfalls - the formation of water users' forums in Kikuletwa catchment, Tanzania. *Water International*, 38(3), 231-249.

Komakech, H. C., Van Der Zaag, P., Mul, M. L., Mwakalukwa, T. A. and Kemerink, J. S. 2012c. Formalization of water allocation systems and impacts on local practices in the Hingilili sub-catchment, Tanzania. *International Journal of River Basin Management*, 10(3), 213-227.

Komakech, H. C., Van der Zaag, P. and Van Koppen, B. 2012d. Dynamics between water asymmetry, inequality and heterogeneity sustain canal institutions in Makanya catchment, Tanzania. *Water Policy*, 14(5), 800-820.

Komakech, H. C., Van der Zaag, P. and Van Koppen, B. 2012e. The last will be first: conflict over water transfers from subsistence irrigation to cities in the Pangani river basin, Tanzania. *Water Alternatives*, 5(3), 700-720.

Komakech, H. C., Van Koppen, B., Mahoo, H. F. and Van der Zaag, P. 2011b. Pangani river basin over time and space: on the interface of local and basin level responses. *Agricultural Water Management*, 98(11), 1740-1751.

Kongo, V. M., 2008. *Balancing water for food and environment: hydrological determinants across scales in the Thukela river basin.* (PhD Thesis). University of KwaZulu-Natal, South Africa.

Kongo, V. M. and Jewitt, G. P. W., Preliminary investigation of catchment hydrology in response to agricultural water use innovations: a case study of the Potshini catchment, South Africa. ed. *In Proceedings of the 11th conference of the South African Council of the Institute of Applied Hydrological Scientists (SANCIAHS)*, September 2005 2005 Midrand.

Kortelainen, J. 1999. The river as an actor-network: The Finnish forest industry utilization of lake and river systems. *Geoforum*, 30(3), 235-247.

Kosgei, J. R., 2009. *Rainwater harvesting systems and their influences on field scale soil hydraulic properties, water fluxes and crop production in Potshini catchment, South Africa*. (PhD Thesis). University of KwaZulu-Natal, South Africa.

Lankford, B. and Beale, T. 2007. Equilibrium and non-equilibrium theories of sustainable water resources management: Dynamic river basin and irrigation behaviour in Tanzania. *Global Environmental Change,* 17(2), 168-180.

Lankford, B. and Hepworth, N. 2010. The cathedral and the bazaar: Monocentric and polycentric river basin management. *Water Alternatives,* 3(1), 82-101.

Lankford, B. and Mwaruvanda, W., 2007. Legal infrastructure framework for catchment apportionment. *In:* Van Koppen, B. and Butterworth, J. A. eds. *Community-Based Water Law and Water Resources Management Reform in Developing Countries.* Wallinford: CABI Publishing.

Lankford, B., Sokile, C. S., Yawson, D. K. and Levite, H., 2004. *The river basin game: a water dialogue tool.* Colombo, Sri Lanka: International Water Management Institute.

Lankford, B., Tumbo, S. D. and Rajabu, K., 2009. Water competition, variability and river basin governance: a critical analysis of the Great Ruaha river, Tanzania. *In:* Molle, F. and Wester, P. eds. *River basin trajectories: Societies, environments and development.* Wallingford, UK and Cambridge, MA, USA: CAB International, 171-195.

Lansing, S. and Miller, J. H. 2005. Cooperation, games, and ecological feedback: some insights from Bali. *Current Anthropology,* 46(2), 328-334.

Latour, B. 1988. Mixing humans and non-humans together: The sociology of a door-closer. *Social Problems,* 35(3), 298-310.

Latour, B., 2005. *Reassembling the social: An introduction to actor-network theory.* Oxford: Oxford University Press.

Law, J. 1992. Notes on the theory of the actor network: Ordering, strategy and heterogeneity. *Systems Practice and Action Research,* 5(4), 379-393.

Lein, H. 2004. Managing the water of Kilimanjaro: Irrigation, peasants, and hydropower development. *GeoJournal,* 61(2), 155-162.

Lein, H. and Tagseth, M. 2009. Tanzanian water policy reforms - between principles and practical applications. *Water Policy,* 11(2), 203-220.

Loeve, R., Hong, L., Dong, B., Mao, G. and Chen, C. D. 2004. Long-term trends in intersectoral water allocation and crop water productivity in Zhanghe and kaifeng, China. *Paddy and Water Environment,* 2(4), 237-245.

Long, N., 1999. The Multiple optic of interface analysis. *UNESCO Background paper on interface analysis.* Wageningen University, The Netherlands.

Long, N., 2001. *Development sociology: actor perspectives.* Routledge, London, UK.

Maganga, F. P. 2003. Incorporating customary laws in implementation of IWRM: some insights from Rufiji River Basin, Tanzania. *Physics and Chemistry of the Earth,* 28(20-27), 995-1000.

Maganga, F. P., Butterworth, J. A. and Moriarty, P. 2002. Domestic water supply, competition for water resources and IWRM in Tanzania: review and discussion paper. *Physics and Chemistry of the Earth,* 27(11-22), 919 - 926.

Maganga, F. P., Kiwasila, H., Juma, I. J. and Butterworth, J. A. 2004. Implications of customary norms and laws for implementing IWRM: Findings from Pangani and Rufiji basins, Tanzania. *Physics and Chemistry of the Earth,* 29(15-18), 1335 - 1342.

Makurira, H., 2010. *Water productivity in rainfed agriculture; redrawing the rainbow of water to achieve food security in rainfed smallholder systems.* (PhD thesis). UNESCO-IHE/TU Delft.

Makurira, H., Mul, M. L., Vygagusa, N. F., Uhlenbrook, S. and Savenije, H. H. G. 2007. Evaluation of community-driven smallholder irrigation in dryland South Pare Mountains, Tanzania: Case study of Manoo micro-dam. *Physics and Chemistry of the Earth,* 32(15-18), 1090 - 1097.

Manzungu, E. 2002. More than a headcount: towards strategic stakeholder representation in catchment management in South Africa and Zimbabwe. *Physics and Chemistry of the Earth, Parts A/B/C,* 27(11-22), 927-933.

Manzungu, E., Senzanje, A. and Van der Zaag, P., eds., 1999. *Water for agriculture in Zimbabwe: policy and management options for the smallholder sector.* Harare: University of Zimbabwe Publications.

Marshall, G. R. 2008. Nesting, subsidiarity, and community-based environmental governance beyond the local level. *International Journal of the Commons,* 2(1), 75-97.

Martin, E. and Yoder, R. 1988. A comparative description of two farmer-managed irrigation systems in Nepal. *Irrigation and Drainage Systems,* 2(2), 147-172.

Masuki, K., 2011. *Farm-level adoption of water system innovations in semi-arid areas: The case of Kakanya watershed in same district, Tanzania.* (PhD Thesis). Sokoine University of Agriculture, Tanzania.

Mbonile, M. J. 2005. Migration and intensification of water conflicts in the Pagani basin, Tanzania. *Habitat International,* 29(1), 41-67.

McCornick, P. G., Awulachew, S. B. and Abebe, M. 2008. Water–food–energy–environment synergies and tradeoffs: major issues and case studies. *Water Policy,* 10(Supplement 1), 23–36.

Mehari, A. H., Van Koppen, B., McCarthy, M. and Lankford, B. 2009. Unchartered innovation? Local reforms of national formal water management in the Mkoji sub-catchment, Tanzania. *Physics and Chemistry of the Earth,* 34, 299 - 308.

Mehta, L. 2007. Whose scarcity? Whose property? The case of water in western India. *Land Use Policy,* 24(4), 654-663.

Mehta, L., Leach, M., Newell, P., Scoones, I., Sivaramakrishnan, K. and Way, S.-A., 1999. *Exploring understandings of institutions and uncertainty: New directions in natural resource management.* Brighton: Institute of Development Studies, University of Sussex.

Meijerink, S. 2008. Explaining continuity and change in international policies: issue linkage, venue change, and learning on policies for the river Scheldt estuary 1967-2005. *Environment and Planning A,* 40(4), 848-866.

Meinzen-Dick, R. and Pradhan, R., 2002. *Legal pluralism and dynamic property rights.* Washington, DC, US: IFPRI.

Meinzen-Dick, R. and Ringler, C. 2008. Water reallocation: drivers, challenges, threats, and solutions for the poor. *Journal of Human Development,* 9(1), 47-64.

Meinzen-Dick, R. and Zwarteveen, M. 1998. Gendered participation in water management: Issues and illustrations from water users' associations in South Asia. *Agriculture and Human Values,* 15(4), 337-345.

Merrey, D. J. 2009. African models for transnational river basin organisations in Africa: an unexplored dimension. *Water Alternatives,* 2(2), 183-204.

Merrey, D. J. and Cook, S. 2012. Fostering institutional creativity at multiple levels: towards facilitated institutional bricolage. *Water Alternatives,* 5(1), 1-19.

Mohamed-Katerere, J. and Van der Zaag, P., 2003. Untying the "knot of silence"; making water policy and law responsive to local normative systems. *In:* Hassan, F. A., Reuss, M., Trottier, J., Bernhardt, C., Wolf, A. T., Mohamed-Katerere, J. and Van der Zaag, P. eds. *History and Future of Shared Water Resources. IHP Technical Documents in Hydrology.* Paris: UNESCO.

Molinas, J. 1998. The impact of inequality, gender, external assistance and social capital on local-level cooperation. *World Development,* 26(3), 413-431.

Molle, F., 2003. *Development trajectories of river basins: A conceptual framework.* Colombo: International Water Management Institute.

Molle, F. 2004. Defining water rights: by prescription or negotiation? *Water Policy,* 6(3), 207-227.

Molle, F. and Berkoff, J., 2006. *Cities versus agriculture: revisiting intersectoral water transfers, potential gains and conflicts.* Colombo Sri Lanka: Comprehensive Assessment Secretariat.

Molle, F. and Berkoff, J. 2009. Cities vs. agriculture: a review of intersectoral water re-allocation. *Natural Resources Forum,* 33(1), 6-18.

Molle, F. and Wester, P., 2009. River basin trajectories: an Inquiry into changing waterscapes. *In:* Molle, F. and Wester, P. eds. *River basin trajectories: societies, environments and development.* CAB International.

Molle, F., Wester, P. and Hirsch, P. 2010. River basin closure: processes, implications and responses. *Agricultural Water Management,* 97(4), 569-577.

Moss, T. 2004. The governance of land use in river basins: prospects for overcoming problems of institutional interplay with the EU Water Framework Directive. *Land Use Policy,* 21(1), 85-94.

Movik, S. 2012. Allocation discourses: South African water rights reform. *Water Policy,* 13(2), 161-177.

Mujwahuzi, M. R., 2001. Water use conflicts in the Pangani basin. *In:* Ngana, J. O. ed. *Water resources management in the Pangani river basin; challenges and opportunities.* Dar es Salaam: Dar es Salaam University Press.

Mul, M. L., 2009. *Understanding hydrological processes in an ungauged catchment in sub-Saharan Africa.* (PhD Thesis). UNESCO-IHE/TU Delft.

Mul, M. L., Kemerink, J., Vygagusa, N. F., Mshana, M. G., van der Zaag, P. and Makurira, H. 2011. Water allocation practices among smallholders farmers in the South Pare Mountains, Tanzania; The issue of scales. *Agricultural Water Management,* 98(11), 1752-1760.

Mul, M. L., Mutiibwa, R. K., Foppen, J. W. A., Uhlenbrook, S. and Savenije, H. H. G. 2007a. Identification of groundwater flow systems using geological mapping and chemical spring analysis in South Pare Mountain, Tanzania. *Physics and Chemistry of the Earth,* 32(15-18), 1015–1022.

Mul, M. L., Mutiibwa, R. K., Foppen, J. W. A., Uhlenbrook, S. and Savenije, H. H. G. 2007b. Identification of groundwater flow systems using geological mapping and chemical spring analysis in South Pare Mountain, Tanzania. *Physics and Chemistry of the Earth,* 32(15 - 18), 1015 - 1022.

Murdoch, J. 1998. The spaces of actor-network theory. *Geoforum,* 29(4), 357-374.

Mwakalukwa, T. A., 2009. *The water allocation systems in Tanzania: understanding the effects of formalization of water allocation systems to indigenous water users in Hingilili catchment.* Thesis (MSc) (MSc). UNESCO-IHE Institute for Water Education.

Naidu, S. C. 2009. Heterogeneity and collective management: Evidence from common forests in Himachal Pradesh, India. *World Development,* 37(3), 676-686.

Neef, A. 2008. Lost in translation: the participatory imperative and local water governance in North Thailand and Southwest Germany. *Water Alternatives,* 1(1), 89 - 110.

Neef, A. 2009. Transforming rural water governance: Towards deliberative and polycentric models? *Water Alternatives,* 2(1), 53-60.

Ngigi, S. N. 2003. What is the limit of up-scaling rainwater harvesting in a river basin? *Physics and Chemistry of the Earth,* 28(20-27), 943 - 956.

Nightingale, A. 2003. A feminist in the forest: Situated knowledges and mixing methods in natural resource management. *ACME: An International E-Journal for Critical Geographies,* 2(1), 77-90.

Olson, M., 1965. *The logic of collective action.* Cambridge, MA: Harvard University Press.

Ostrom, E., 1990. *Governing the commons: The evolution of institutions for collective action.* New York: Cambridge University Press.

Ostrom, E. 1993. Design principles in long-enduring irrigation institutions. *Water Resources Research,* 29(7), 1907-1912.

Ostrom, E., 1998. Self-governance and forest resources. *International CBNRM Workshop.* Washington, DC, US.

Ostrom, E. 1999. Coping with tragedies of the commons. *Annual Review of Political Science,* 2(1), 493-535.

Ostrom, E. 2000. Collective action and the evolution of social norms. *Journal of Economic Perspectives,* 14(3), 137-158.

Ostrom, E., 2002. Chapter 24 Common-pool resources and institutions: Toward a revised theory. *In:* Bruce, L. G. and Gordon, C. R. eds. *Handbook of Agricultural Economics.* Elsevier, 1315-1339.

Ostrom, E. 2010. Polycentric systems for coping with collective action and global environmental change. *Global Environmental Change,* 20, 550-557.

Ostrom, E., Burger, J., Field, C. B., Norgaard, R. B. and Policansky, D. 1999. Revisiting the commons: Local lessons, global challenges. *Science,* 284, 278-282.

Ostrom, E. and Gardner, R. 1993. Coping with asymmetries in the commons: Self-governing irrigation systems can work. *Economic Perspectives,* 7(4), 93-112.

Ostrom, E., Gardner, R. and Walker, J., 1994. *Rules, games and common-pool resources.* Michigan: The University of Michigan Press.

Ostrom, V., Tiebout, C. and Warren, R. 1961. The organization of government in metropolitan areas. *American Political Science Review,* 55, 831-842.

Pachpute, J. S., Tumbo, S. D., Sally, H. and Mul, M. L. 2009. Sustainability of Rainwater Harvesting Systems in Rural Catchment of Sub-Saharan Africa. *Water Resources Management,* 23(13), 2815-2839.

PAMOJA, 2004. Dialogue on water: situation brief. Moshi.

Pamoja, 2006. *Organisational landscape: Kikuletwa subcatchment, Pangani basin.* Moshi: PAMOJA Trust.

PBWO/IUCN, 2007. *Pangani river system: state of the basin report 2007.* Nairobi PBWO Moshi, Tanzania and IUCN Eastern Africa Regional Program.

Perramond, E. P. 2012. The politics of scaling water governance and adjudication in New Mexico. *Water Alternatives,* 5(1), 62-82.

Piccione, M. and Razin, R. 2009. Coalition formation under power relations. *Theoretical Economics,* 4(1), 1-15.

Pierson, P. 2000. Increasing returns, path dependence, and the study of politics. *American Political Science Review,* 94(2), 251-267.

Poncelet, E. C. 2001. A kiss here and a kiss there: Conflict and collaboration in environmental partnerships. *Environmental Management,* 27(1), 13-25.

Poteete, A. R. and Ostrom, E. 2004. Heterogeneity, group size and collective action: The role of institutions in forest management. *Development and Change,* 35(3), 435-461.

Potkanski, T. and Adams, W. M. 1998. Water scarcity, property regimes and irrigation management in Sonjo, Tanzania. *The Journal of Development Studies,* 34(4), 86-116.

Quinn, C. H., Huby, M., Kiwasila, H. and Lovett, J. C. 2007. Design principles and common pool resource management: An institutional approach to evaluating community management in semi-arid Tanzania. *Journal of Environmental Management,* 84(1), 100-113.

Robinson, D. J. and Smith, M., eds., 2010. *Negotiate - Reaching agreements over water.* Gland, Switzerland: IUCN.

Rockström, J., Folke, C., Gordon, L. J., Hatibu, N., Jewitt, G. P. W., Penning de Vries, F., Rwehumbiza, F., Sally, H., Savenije, H. H. G. and Schulze, R. 2004. A watershed approach to upgrade rainfed agriculture in water scarce regions through water system innovations: an integrated research initiative on water for food and rural livelihoods in balance with ecosystem functions. *Physics and Chemistry of the Earth,* 29(15-18), 110 - 1118.

Rosegrant, M. W., Ringler, C. and Zhu, T. 2009. Water for agriculture: maintaining food security under growing scarcity. *The Annual Review of Environment and Resources,* 34, 205-222.

Runge, C. F. 1986. Common property and collective action in economic development. *Water Development,* 14(5), 623 - 635.

Ruthenberg, H., 1980. *Farming Systems in the Tropics.* Oxford: Clarendon Press.

Ruttan, L. M. 2008. Economic heterogeneity and the commons: Effects on collective action and collective goods provisioning. *World Development,* 36(5), 969-985.

Sääksvuori, L., Mappes, T. and Puurtinen, M. 2011. Costly punishment prevails in intergroup conflict. *Proceedings of the Royal Society B: Biological Sciences,* 278(1723), 3428-3436.

Sadiki, H., 2008. Water resources management challenges in Pangani basin. *Ministry of Water and Irrigation - World Bank Institute cum Training on IWRM.* Arusha: Pangani Basin Water Office.

Same District report, 2008. *Halmashauri ya wilaya ya Same: Idadi ya watu katika vijiji kata mwaka 2008 eneo la utawala na idadi ya watu (official).* Same District, Tanzania.

Sarker, A. and Itoh, T. 2001. Design principles in long-enduring institutions of Japanese irrigation common-pool resources. *Agricultural Water Management,* 48(2), 89-102.

Sarmett, J., Burra, R., van Klinken, R. and Kelly, W., 2005. Managing water conflict through dialogue in Pangani basin, Tanzania. *FAO/Netherlands International conference on Water for food and ecosystems.* The Hague: FAO.

Savenije, H. H. G. and Van der Zaag, P. 2002. Water as an economic good and demand management; paradigms with pitfall. *Water International,* 27(1), 98-104.

Schlager, E. and Ostrom, E. 1992. Property-rights regimes and natural resources: a conceptual analysis. *Land Economics,* 68(3), 249-262.

Sehring, J. 2009. Path dependencies and institutional bricolage in post-soviet water governance. *Water Alternatives,* 2(1), 61-81.

Sheridan, M., J 2002. An irrigation intake is like a uterus: culture and agriculture in precolonial North Pare, Tanzania. *American Anthropologist,* 104(1), 79 - 92.

Shi, T. 2006. Simplifying complexity: raitionalizing water entitlements in the southern connected river Murray system, Australia. *Agricultural Water Management,* 86(3), 229-239.

Sokile, C. S., Kashaigili, J. J. and Kadigi, R. M. J. 2003. Towards an integrated water resource management in Tanzania: the role of appropriate institutional framework in Rufiji Basin. *Physics and Chemistry of the Earth,* 28(20-27), 1015 - 1023.

Sokile, C. S., Mwaruvanda, W. and Van Koppen, B., Integrated water resources management in Tanzania: interface between formal and informal institutions. ed. *Paper presented in the International workshop on 'African Water Laws: Plural Legislative Frameworks for Rural Water Management in Africa', 26-28 January 2005,* 2005 Johannesburg, South Africa.

Sokile, C. S. and Van Koppen, B. 2004. Local water rights and local water user entities: the unsung heroines of water resource management in Tanzania. *Physics and Chemistry of the Earth,* 29(15-18), 1349 - 1356.

Sotthewes, W., 2008. *Forcing on the salinity distribution in the Pangani Estuary.* (MSc. Thesis). Delft University of Technology.

Sovacool, B. K. 2011. An international comparison of four polycentric approaches to climate and energy governance. *Energy Policy,* 39(6), 3832-3844.

Spear, T., 1997. *Mountain farmers: Moral economies of land and agricultural development in Arusha and Meru.* Oxford: James Currey.

Swatuk, L. A. 2008. A Political economy of water in Southern Africa. *Water Alternatives,* 1(1), 24-47.

Swyngedouw, E. 1997. Power, nature, and the city. the conquest of water and the politcal ecology of urbanization in Guayaquil, Ecuador: 1880-1990. *Environment and Planning A,* 29(2), 311-332.

Syme, G. J., Nancarrow, B. E. and McCreddin, J. A. 1999. Defining the components of fairness in the allocation of water to environmental and human uses. *Journal of Environmental Management,* 282(57), 51-70.

Tack, S. P., 2006. *Drowning in conflict floating on dialogue: The influence of geographical capital in water management in semi-arid Tanzania.* (MSc thesis). Utrecht University.

Tagseth, M. 2008. Oral history and the development of indigenous irrigation. Methods and examples from Kilimanjaro, Tanzania. *Norsk Geografisk Tidsskrift - Norwegian Journal of Geography,* 62(1), 9 - 22.

Tanzania, 2002a. *National Water Policy* Dar es Salaam: Ministry of Water and Livestock Development.

Tanzania, 2002b. *Population and housing census.* Dar es Salaam: Government Printer.

Tanzania, 2009. *The water resources management act supplement No. 11.* Government Printer, Dar es Salaan.

Tarimo, A. K. P. R., Mdoe, N. S. and Lutatina, J. M. 1998. Irrigation water prices for farmer-managed irrigation systems in Tanzania: a case study of Lower Moshi irrigation scheme. *Agricultural Water Management*, 38(1), 33 - 44.

Territory, T. U. T., 1948. *Water Ordinance of 1948*. Dar-es-Salaam: Government Printers.

Tumbo, S. D., Mutabazi, K. D., Byakugila, M. M. and Mahoo, H. F. 2011. An empirical framework for scaling-out of water system innovations: lessons from diffusion of water system innovations in the Makanya catchment in Northern Tanzania. *Agricultural Water Management*, 98(11), 1761-1773.

Turpie, J., Ngaga, Y. and Karanja, F., 2003. *A preliminary economic assessment of water resources of the Pangani river basin, Tanzania: Economic value, incentives for sustainable use and mechanisms for financing management.* . Nairobi: IUCN.

Turpie, J., Ngaga, Y. and Karanja, F., 2004. *Maximising the economic value of water resources in the Pangani river basin, Tanzania.* . Nairobi: IUCN - The World Conservation Union, Eastern Africa Region office.

Van der Zaag, P. 2007. Asymmetry and equity in water resources management: Critical institutional issues for Southern Africa. *Water Resources Management*, 21(12), 1993-2004.

Van der Zaag, P. and Röling, N., 1996. The water acts in the Nyachowa catchment area. *In:* Manzungu, E. and Van der Zaag, P. eds. *The practice of smallholder irrigation: case studies from Zimbabwe.* Harare: University of Zimbabwe Publications, 161-190.

Van Koppen, B. 2003. Water reform in Sub-Saharan Africa: what is the difference? *Physics and Chemistry of the Earth*, 28(20-27), 1047 - 1053.

Van Koppen, B., Shah, T., Namara, R., Barry, B. and Van der Zaag, P., 2008. Water rights in informal economies in the Limpopo and Volta basins.

Van Koppen, B., Sokile, C. S., Hatibu, N., Lankford, B., Mahoo, H. F. and Yanda, P. Z., 2004. *Formal water rights in rural Tanzania: Deepening the dichotomy?* . Colombo: International Water Management Institute.

Van Koppen, B., Sokile, C. S., Lankford, B., Hatibu, N., Mahoo, H. F. and Yanda, P. Z., 2007. Water rights and water fees in rural Tanzania. *In:* Molle, F. and Berkoff, J. eds. *Irrigation water Pricing.* CAB International, 143 - 163.

Varughese, G. and Ostrom, E. 2001. The contested role of heterogeneity in collective action: Some evidence from community forestry in Nepal. *World Development*, 29(5), 747-765.

Vavrus, F. K. 2003. A Shadow of the real thing": Furrow Societies, Water User Associations, and Democratic Practices in the Kilimanjaro Region of Tanzania. *The Journal of African American History*, 88(4), 393 - 412.

Vedeld, T. 2000. Village politics: heterogeneity, leadership and collective action. *The Journal of Development Studies*, 36(5), 105-134.

Waalewijn, P., Wester, P. and van Straaten, K. 2005. Transforming river basin management in South Africa; Lessons from the lower Komati river. *Water International*, 30(2), 184-196.

Wade, R., 1988. *Village republics: Economic conditions for collective action in South India.* New York: Cambridge University press, USA.

Wang, L. Z., Fang, L. and Hipel, K. W. 2003. Water resources allocation: A cooperative game theoretic approach. *Journal of Environmental Informatics*, 2(2), 11-22.

Warner, J. 2005. Multi-stakeholder platform: integrating society into integrated water resources management? *Ambient and sociedade*, 8(2), 9 - 28.

Warner, J., Wester, P. and Bolding, A. 2008. Going with the flow: river basins as the natural units for water management? *Water Policy,* 10(Supplement 2), 121-138.

Wester, P., Merrey, D. J. and De Lange, M. 2003. Boundaries of consent: stakeholder representation in river basin management in Mexico and South Africa. *World Development,* 31(5), 797-812.

Wester, P., Vargas-Velázquez, S., Mollard, E. and Silva-Ochoa, P. 2008. Negotiating Surface Water Allocations to Achieve a Soft Landing in the Closed Lerma-Chapala Basin, Mexico. *International Journal of Water Resources Development,* 24(2), 275-288.

World Bank, 1996. *Staff appraisal report, Tanzania.* Washington.

Yin, R. K., 2003. *Case study research: Design and methods.* 3 ed. London, UK: Sage Publication.

Young, O., R, 2003. *The institutional dimensions of environmental change: fit, interplay, and scale.* Cambridge, Massachusetts, USA: MIT Press.

Zwarteveen, M., Roth, D. and Boelens, R., 2005. Water rights and legal pluralism: Beyond analysis and recognition. *In:* Roth, D., Boelens, R. and Zwarteveen, M. eds. *Liquid relations. Contested water rights and legal complexity.* New Brunswick, NJ: Rutgers University press, 254 -268.

SAMENVATTING

In veel stroomgebieden in sub-Sahara Afrika en in andere delen van de wereld nemen waterbeheer problemen toe door toedoen van snelle verstedelijking, armoede en voedselschaarste in midden- en lage lonen landen, evenals vanwege de toenemende vraag naar energie en klimaat verandering. Bijna de helft van de wereldbevolking leeft in steden en geschat wordt dat dit groeit naar tweederde in 2050. De toenemende vraag naar water in stedelijke gebieden brengt nieuwe uitdagingen voor rivierbeheer met zich mee. Herverdeling van water van andere sectoren naar steden is een voor de hand liggende oplossing, maar dit kan verstrekkende gevolgen hebben in een stroomgebied. Daarbij neemt rurale armoede, honger en voedselzekerheid in sub-Sahara Afrika toe. Om de toename van rurale armoede te stoppen en/of ongedaan te maken en om werkgelegenheid te genereren zijn aanzienlijke investeringen in de geïrrigeerde landbouw nodig. Hervorming van de landbouw in sub-Sahara Afrika betekent echter ook ingrijpen in waterbeheer, aangezien gebrek aan betrouwbare watertoevoer één van de grootste beperkingen is voor gewasproductie. Door de bovengenoemde problemen en de wereldwijd stijgende voedsel- en energieprijzen worden buitenlandse investeringen aangetrokken. Directe buitenlandse investeringen in de landbouw van sub-Sahara Afrika zullen waarschijnlijk het agrarisch watergebruik vergroten en dit kan een reeds nijpende watersituatie verslechteren.

Op veel plaatsen proberen zowel de watergebruikers als de overheid in te gaan op een aantal van de bovengenoemde uitdagingen, bijvoorbeeld door meer water uit de rivier te onttrekken, door het bouwen van reservoirs of door te zoeken naar alternatieve bronnen van water zoals grondwater. Deze interventies kunnen echter watertekorten veroorzaken en daarmee gebruikers elders in het stroomgebied benadelen. Een toename van watertekorten leidt tot concurrentie en conflicten tussen gebruikers, groot en kleine, boven- en benedenstrooms. Door de toenemende concurrentie rondom watergebruik worden bestaande endogene water instituties en hun vermogen om oplossingen te vinden voor strijdige belangen zwaarder belast. Naast het nemen van maatregelen die de waterbeschikbaarheid vergroten, vereist het oplossen van conflicten op het gebied van water nieuwe bestuurlijke structuren die een rechtvaardig en duurzaam gebruik van de beperkte hoeveelheid water garanderen. Dit betreft ook het aanpassen van regelgeving rondom het verdelen van water tussen concurrerende gebruikers en het aanpassen van organisatiestructuren betrokken bij het toezicht op, en naleving van, de waterverdelingafspraken. Het begrijpen van institutionele veranderingsprocessen en implementatiemethodes staan daarom centraal bij het oplossen van de waterbeheeruitdagingen waar samenlevingen in waterschaarse stroomgebieden over de hele wereld voor staan.

Veel overheden in sub-Sahara Afrika hebben nieuw beleid en wetten gemaakt en nieuwe instituties gecreëerd om rechtvaardig en duurzaam waterbeheer te bevorderen. Het formaliseren van het eigendomsrecht op water en participatie van gebruikers in zogenoemde stroomgebiedfora worden geacht de coördinatie en het oplossen van water conflicten te verbeteren. Echter, overheidsinterventies in waterbeheer en formele regelgeving nemen vaak endogene waterbeheerspraktijken niet serieus. Sommige endogene waterbeheerpraktijken zijn goed bekend, zeker in (semi-)aride gebieden waar ze uitgegroeiden tot succesvolle instituties voor het delen van water. Deze lokaal geëvolueerde instituties, mits goed begrepen, kunnen een vervanging zijn voor - of gebruikt worden ter verbetering van - waterbeheerinstituties die door veel overheden ingevoerd worden op stroomgebiedniveau. De uitdaging is dat deze aanpak lokale praktijken wil opschalen, terwijl staatsinstituties tegelijkertijd worden aangepast aan lokale condities. Dit vereist tevens begrip waarom endogene instituties ontstaan, hoe ze functioneren, voortduren en op welk schaalniveau ze effectief kunnen blijven.

Dit proefschrift draagt bij aan deze problematiek door een Afrikaans stroomgebied te bestuderen, namelijk de Pangani in Tanzania. Het stroomgebied is een perfect levend laboratorium om het ontstaan en ontwikkeling te bestuderen van lokale en staats waterbeheerinstituties. De Pangani is een gedeeltelijk gesloten stroomgebied. Gedeeltelijk omdat sommige zijtakken droogvallen gedurende periodes in het jaar door teveel watergebruik. Het is gedeeltelijk open omdat het gebruik van grondwater nog onderontwikkeld is; er is erg weinig bekend over grondwater gebruik, beschikbaarheid en de relatie met het oppervlaktewater. Het is een stroomgebied waar staats ingrijpen terug te voeren is tot de koloniale tijd en er zich al meer dan 100 jaar lokale gebruiken ontwikkelen. Het overkoepelende onderzoeksdoel was om de voorwaarden te onderzoeken waarbij institutionele arrangementen van de overheid samen kunnen gaan met lokale waterbeheerpraktijken. Dit proefschrift is gebaseerd op bevindingen van meerdere gevalstudies in de Pangani in Tanzania. Met actoren zijn diepte interviews en rollenspellen gehouden, gevolgd door een terugkoppelingsworkshops, en geïnformeerd door zorgvuldig in kaart gebrachte irrigatiekanalen en geïrrigeerde landbouwpercelen en gebieden.

De bevindingen van dit proefschrift geven aan dat in plaats van harmonie, de overheidsingrepen in de watersector resulteren in dissonantie met de lokaal ontwikkelde water instellingen. In de Pangani wordt de staatgestuurde formalisering van het recht op het bezit van water gebruikt door nieuwe gebruikers om toegang en controle te krijgen over water ten koste van de bestaande gebruikers. Waterrechten zoals toegepast in de Pangani zijn moeilijk af te dwingen en te controleren, en tot nu toe heeft dit ook niet geleid tot efficiënt watergebruik. Betekenisvolle participatie van watergebruikers in het besluitvormingsproces rondom het stroomgebiedbeheer blijkt problematisch. In een deelstroomgebied, de Kikuletwa, is bewezen dat het moeilijk is om de meest geschikte hydrologische beheerseenheid te vinden voor besluitvorming, die ook aansluit bij de bestaande politiek-administratieve eenheid. De manier waarop de institutionele verweving in de Kikuletwa werd opgezet werkte niet. Het opdelen van het grotere Kikuletwa deelstroomgebied in kleinere delen met elk eigen verenigingen voor watergebruikers creëerde enkel extra beheerslagen zonder de noodzakelijke integratie van bestaande, lokaal ontwikkelde, arrangementen, zoals bijvoorbeeld rivier comités. De nieuw gevormde Kikuletwa substroomgebied watergebruikersverenigingen zijn eerder een soort eilanden die niet goed geïntegreerd zijn in de bestaande arrangementen. Watergebruikers zien niet hoe ze de deelstroomgebied watergebruikersverenigingen in hun eigen bestuurlijke regelingen kunnen inpassen. De algemene conclusie met betrekking tot staatsingrijpen in waterbeheer is dat het probleem van het inpassen van verschillende soorten instituties ("institutional fit') en tegelijkertijd het integreren van informele afspraken in de staatsgestuurde bestuurlijke structuren een zorgvuldige analyse vereist, met meeneming van de sterke en zwakke punten, van de bestaande lokale structuren.

Ook al wordt er algemeen gedacht dat het toewijzen van waterrechten of vergunningen in waterschaarse stroomgebieden leidt tot verbetering van gelijkheid en vermindering van conflicten, tonen de bevindingen in dit proefschrift aan dat de 'papieren' waterrechten mogelijk gebruikt worden door nieuwe gebruikers om toegang te krijgen tot water, ten koste van oude gebruikers. Het waterrechten systeem zoals toegepast door de Tanzaniaanse overheid in the Pangani maakt het juridisch mogelijk dat machtige spelers bestaande gebruikers onteigenen. Steden in de Pangani passen de wet selectief toe om meer controle te krijgen over water. In andere gevallen wordt de legitimiteit van het staatsgestuurde waterrechtensysteem in twijfel getrokken door verschillende partijen. In de Pangani, maken kleine gebruikers aanspraak op het gewoonterecht, terwijl grootschalige irrigatie gebruikers toegang tot water proberen te krijgen door aanspraak te maken op de formele waterwet. Alhoewel de meeste grootschalige (commerciële) boeren voordeel hebben door hun locatie, wordt hun 'officiële waterrecht' uitgedaagd door de benedenstroomse kleinschalige boeren. Deze boeren eisen dat verdeling van het water op basis van rotatie gebeurt en dat variatie in levering mogelijk is in plaats van het toepassen van de absolute volumes zoals gespecificeerd in het formele waterrecht.

Dit proefschrift toont aan dat vernieuwing van institutionele regelingen op lokaal niveau met betrekking tot het verdelen van water in het verleden vaak ontstond door het creëren van hydraulisch eigendom en/of werd onderhandeld om meer afvoer voor benedenstrooms te garanderen. De hydraulische locatie van de verschillende gebruikers in het stroomgebied (bovenstrooms of benedenstrooms) is de belangrijkste drijfveer voor institutionele innovatie. In de bestudeerde casussen waren het altijd de benedenstroomse gebruikers die het proces van institutionele verandering in het stroomgebied initieerden. In tegenstelling tot veel onderzoek naar collectieve actie waar water asymmetrie, ongelijkheid en heterogeniteit gezien wordt als risico voor collectieve actie, toont dit proefschrift aan dat deze dynamisch op elkaar inwerken en aanleiding geven tot onderlinge afhankelijkheden tussen watergebruikers wat leidt tot coördinatie en samenwerking. De resultaten met betrekking tot collectieve actie beperkt zich tot relatief kleine ruimtelijke en sociale schaal; meestal betreft het boeren van één dorp. In dergelijke situaties kunnen er belemmeringen zijn voor unilaterale acties door sociale- en groepsdruk. Nabijheid kan dus een noodzakelijke voorwaarde zijn voor het ontstaan van collectieve actie in asymmetrische verdelingssituaties. Op grotere ruimtelijke schalen en over grotere afstanden, bijvoorbeeld deelstroomgebieden of hele stroomgebieden, zal dit allicht anders zijn. De grootste ruimtelijke schaal waarop lokale watergebruikers het water verdeelden was een stuk van de rivier van drie administratieve wards (circa 15 km). Geen lokaal ontstane regeling is gevonden buiten deze ruimtelijke schaal. Dit kan zijn doordat buiten de kleine ruimtelijke schaal, het creëren en behouden van collectieve actie moeilijk is.

Dit proefschrift draagt bij aan bestaande theorieën en concepten op het gebied van watermanagement in een stroomgebied. Het proefschrift heeft Molle's (2003) typologie van reacties van stroomgebiedsactoren uitgebreid door expliciet een meso laag toe te voegen welke het raakvlak is van staatsgestuurde en lokale initiatieven. Het laat ook zien dat niet alle acht institutionele ontwerpprincipes van Ostrom (1993) noodzakelijk zijn om waterinstituties effectief te laten zijn en te blijven. Het proefschrift geeft ook een conceptuele verklaring voor de dynamiek tussen water asymmetrie, ongelijkheid in toegang tot land en heterogeniteit voor behoud van collectieve actie met betrekking tot gemeenschappelijke waterbronnen.

De hiërarchische structuur van de staatsinstitutie die lokale waterbeheerarrangementen inbedde werkte niet in het bestudeerde stroomgebied; gedeeltelijk vanwege de manier waarop het was toegepast, maar ook vanwege de complexe overlappende jurisdictie tussen staat- en lokaalontwikkelde instellingen. Er zijn echter mogelijkheden om de staatsgestuurde stroomgebiedbeheer structuur te integreren met lokale waterbeheers arrangementen. Wij vonden dat in de Pangani de rivier comités de meest veelbelovende lokaal ontwikkelde instellingen waren die de staatsgestuurde en lokaal ontstane water regelingen konden verzoenen. Eén beleidsaanbeveling is dat een rivier comité collectieve waterrechten zou kunnen uitgeven met het mandaat om een minimale afvoer voor het gebied benedenstrooms van zijn jurisdictie te garanderen. Op deze manier zouden de water organisaties in het stroomgebied geen waterrechten uitgeven welke ze niet kunnen handhaven en monitoren. In plaats daarvan zullen ze beperkte middelen inzetten om naleving van de waterrechten te bewaken.

Echter, meer onderzoek is nodig om de rol van de gemeentelijke overheid in het aanpakken van de watercompetitie op grotere ruimtelijke schalen te begrijpen. Dit proefschrift heeft niet de dynamiek van sekseongelijkheid en toegang tot water in detail bestudeerd. Leiderschap van lokale en staatsgestuurde waterbeheer organisaties in de Pangani wordt gedomineerd door mannen en in zulke situaties kan de gelijkheid en rechtvaardigheid voor vrouwen verslechteren. Een betere combinatie van staatsgestuurde en lokale waterbeheerinstituties met betere controles op de meer machtige actoren kunnen zowel gunstig zijn voor vrouwen als andere gemarginaliseerde groepen. Dit vereist vervolgonderzoek. In tegenstelling tot de Pangani, zijn de meeste stroomgebieden in sub-Sahara Afrika nog steeds open. In deze stroomgebieden kan het vergroten van het wateraanbod door alternatieve bronnen nog steeds de eerste stap zijn. Echter, sinds het creëren van hydraulisch eigendom ook de relatie tussen de actoren verandert is vervolgonderzoek nodig naar het vergelijken van de hydraulische eigendomsrechten (infrastructurele ontwikkelingen om het wateraanbod te vergroten) met arrangementen die gericht zijn om het schaarse water eerlijk te verdelen. Om meer inzicht te krijgen in het functioneren van zelfbestuur is tenslotte aanvullend onderzoek nodig: 1) beschrijf de verschijnselen van water asymmetrie, ongelijkheid en heterogeniteit op grotere ruimtelijke schalen en analyseer de omstandigheden onder welke deze plaatsvinden; en 2) verifieer de relatie tussen ongelijkheid met betrekking tot toegang tot land en water in open kanaal systemen en het collectieve vermogen om water te delen en arbeid te mobiliseren voor onderhoud aan vele andere open kanaal systemen, dit laatste om de bevindingen van dit proefschrift te generaliseren, niet alleen in de Pangani, maar ook in de Rufiji in Tanzania en andere Afrikaanse landen zoals Kenia en Mozambique, en wellicht ook in andere continenten zoals in Nepal.

ABOUT THE AUTHOR

Charles Hans Komakech, born in 1975 in Kitgum, Uganda, obtained his bachelor degree in Civil Engineering from Makerere University, Kampala. After graduating, Hans worked in several engineering construction projects both within and outside of Uganda. In 2003, Hans was awarded the Netherlands Fellowship scholarship to study at UNESCO-IHE. He obtained his Master of Science degree in Water Management (specialisation of Water Resources Management). Hans was also awarded the Commonwealth Commission Scholarship and in 2008 he obtained a second Master of Science degree in Water and Waste Engineering from Loughborough University, UK.

In 2007, Hans started his PhD research in the Pangani river basin, Tanzania. The research was part of a larger multidisciplinary programme called the Smallholder System Innovations in Integrated Watershed Management (SSI) programme. The focus of Hans PhD was on understanding the emergence and evolution of water institutions in the Pangani basin, Tanzania. In addition to researching the effect of state-led interventions, Hans also studied the dynamics impacts of water asymmetry, inequality and heterogeneity on collective action institutions. He also supervised Master students and presented his work at several international conferences.

In 2012, Hans joined the department of Water, Environmental Science and Engineering, at the Nelson Mandela African Institute of Science and Technology, in Arusha, Tanzania, as a lecturer of integrated watershed and river basin management. His research interests include understanding the emergence of collective action institutions, water allocation and governance, participatory simulation and agent-based modelling, and agricultural water management.